Marina Olsson
Park Min-Jae

Long COVID

A medical guide for those affected and their relatives

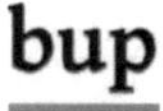

Marina Olsson
Park Min-Jae

Long COVID

A medical guide for those affected and their relatives

ISBN: 978-3-68904-432-9
Order number: 1400 (Paperback)
Also available as an eBook

Bremen University Press, 2024.

First edition
May 2024
Printed in the European Union
bup@bremenuniversitypress.com
www.bremenuniversitypress.com

Marina Olsson
Park Min-Jae

Long COVID

A medical guide for those affected and their relatives

Overview

Table of contents

C. COMMON SYMPTOMS OF LONG COVID AND THEIR EFFECTS   51

D. NEUROLOGICAL AND COGNITIVE SYMPTOMS   73

A. Introduction

In recent years, the global health community has faced an unprecedented challenge: the COVID-19 pandemic. While much of the research and public attention has focused on the acute aspects of the disease, a new, complex phenomenon has now come to the fore: Long COVID.

This book is dedicated to this phenomenon, which is characterised by a variety of long-lasting symptoms that persist after the initial recovery from SARS-CoV-2 infection.

Long COVID encompasses a broad spectrum of clinical manifestations, ranging from persistent fatigue and respiratory problems to neurological and cardiovascular disorders. This disease poses a significant challenge not only for patients, but also for physicians, researchers and healthcare systems worldwide.

The aim of this textbook is to provide knowledge and understanding of Long COVID, including the epidemiological, physiological and clinical aspects. It offers current research findings, evidence-based treatment approaches and interdisciplinary management strategies. It is primarily aimed at those affected, relatives and an interested specialist audience.

II Definition of long COVID and differentiation from acute COVID-19

Long COVID, also known as post-COVID-19 syndrome, refers to a condition in which symptoms persist after an initial COVID-19 infection or new symptoms appear after recovery. This condition can manifest when symptoms persist for weeks or months and significantly affect the quality of life of those affected. Long COVID is not limited to those who have had a severe acute illness; individuals with initially mild symptoms can also be affected.

In contrast, acute COVID-19 refers to the phase of direct viral infection, in which symptoms typically appear within two weeks of infection and usually resolve completely within a few weeks. The acute phase is characterised by symptoms such as fever, cough and breathing difficulties and can lead to hospitalisation in severe cases.

Long COVID includes a wide range of symptoms such as fatigue, breathing problems, chest pain, joint and muscle pain, neurological disorders, cognitive impairment ("brain fog"), mood swings and others. These symptoms of Long COVID can occur intermittently or continuously over long periods of time and often have an intermittent course, with symptoms improving or worsening at times.

The distinction between acute COVID-19 and long COVID is important as it requires different treatment and management approaches. While the treatment of acute COVID-19 often focuses on the immediate reduction of viral load and treatment of respiratory symptoms, the treatment of long COVID focuses on a multidisciplinary approach to alleviate persistent symptoms and improve the quality of life of those affected.

III Significance of the topic in the context of the global health situation

The topic of Long COVID has significant importance in the context of the global health situation, especially as the COVID-19 pandemic has infected millions of people worldwide. The long-term consequences of this infection raise important questions for the healthcare system, the economy and society in general.

Firstly, Long COVID places a considerable burden on healthcare systems. Many of those affected require long-term medical care, ranging from general medical treatment to specialised therapies. This increases the pressure on already strained healthcare systems, which also have to deal with acute cases of COVID-19 and other medical needs.

Secondly, Long COVID has economic consequences. People who suffer from long-lasting symptoms are often unable to work or are limited in their ability to work, leading to absenteeism and reduced productivity. This has far-reaching effects on the economy, including increased costs for healthcare and social benefits.

Thirdly, Long COVID influences public perception and management of COVID-19, emphasising the need to remain vigilant even after the acute phase of the pandemic has subsided. The continued presence of Long COVID requires public health strategies that go beyond acute infection control to include rehabilitation and support services for those affected.

Fourthly, Long COVID has social and psychological effects. Many people experience social isolation or depression due to their symptoms. This emphasises the need for psychosocial support and awareness of Long COVID conditions.

Overall, the topic of Long COVID shows how important it is to look at pandemic diseases holistically, including the long-term effects they can have on individuals and societies. Given the global spread of COVID-19 and the number of people who could potentially suffer long-term consequences, Long COVID remains a key topic in the global health discussion.

IV Statistical data on distribution and the demographics affected

The statistics on the spread of Long COVID and the demographics affected vary by study and region, but provide an insight into the scale and complexity of the problem. Some key information and trends can be derived from various international studies

a. Spread of long COVID

The data on the spread of Long COVID varies widely, largely due to the different methods of data collection, the timing of the follow-up and the different definitions of the condition.

Studies and estimates on Long COVID show that around 10% to over 30% of people who become infected with the coronavirus later develop persistent or new symptoms.

The wide variation in percentages may also be influenced by demographic factors, such as the age of those affected, pre-existing medical conditions and even the severity of the original COVID-19 disease. In addition, the quality of healthcare systems and access to medical care play a role in how well and how early Long COVID is recognised and treated.

These uncertainties in the data make it difficult to make accurate predictions about the long-term impact of Long COVID on society and the healthcare system. Intensive research is therefore continuing in order to develop better diagnostic procedures, better understand the various symptoms and their progression and ultimately find more effective treatment methods.

b. Affected groups

Long COVID shows a clear variation in the extent to which different demographic groups are affected. While the syndrome can generally affect people of all ages, there are certain groups that have a higher prevalence.

Middle-aged and older adults are more commonly affected by Long COVID. This could be partly because older people are generally at higher risk of more severe courses of COVID-19 and may therefore be more susceptible to lingering after-effects of the disease. The higher prevalence in this group could also be due to a pre-existing burden of other chronic health conditions that could be exacerbated by COVID-19.

Interestingly, there is also a striking gender difference in long COVID: women report long-lasting symptoms disproportionately more frequently than men.

This could be due to several factors. Firstly, biological differences, such as immune response and hormonal factors, could play a role. Women tend to have a stronger immune response, which protects them from many infections, but also leads to a higher rate of autoimmune diseases, which may influence the long-term symptoms of COVID-19. Secondly, social and behavioural factors could play a role, such as the different roles that men and women play in society and in health behaviour.

Children and adolescents are not immune to Long COVID either, although the symptoms in these groups tend to be less frequent and often less severe than in adults. Nevertheless, it is important to be aware of the long-term consequences of COVID-19 in younger age groups, especially because they can affect growth and development.

The exact mechanisms that lead to these differences in susceptibility and symptoms of Long COVID are the subject of current research. This knowledge is crucial in order to develop targeted treatment strategies that are tailored to the needs of specific population groups.

c. Severity of the original COVID-19 disease

The severity of the initial COVID-19 illness plays a recognisable role in the development of Long COVID, however, the relationship between the severity of the initial infection and the subsequent onset of long-term symptoms is complex and not completely linear. Individuals who experienced more severe symptoms during their initial COVID-19 illness, especially those who were hospitalised or required oxygen support, actually show a higher risk of suffering from Long COVID. This observation can be partly explained by the fact that a more severe infection could trigger a more intense and prolonged immune system response, which in turn leads to a greater likelihood of persistent inflammatory processes and resulting symptoms.

Interestingly, however, it can also be seen that even people who originally showed only mild symptoms of

COVID-19 can develop Long COVID. This suggests that the pathogenesis of Long COVID is not solely dependent on the initial severity of the disease. The reasons for this are manifold and could include genetic predispositions, individual differences in the immune system and possibly previously unknown viral or immunological factors.

These findings highlight the need to consider Long COVID as a multifaceted syndrome that is influenced by a combination of biological, immunological and possibly social factors. Researchers are endeavouring to further decipher the mechanisms that lead to Long COVID in order to improve both preventative measures and therapeutic approaches that target not only those who were severely ill, but also those who initially had milder courses.

d. Socio-economic and ethnic factors

The observation of socioeconomic and ethnic disparities in the incidence of Long COVID is another important aspect that shows how societal inequalities can impact health outcomes. Indeed, in the US and UK, studies have shown that ethnic minorities and those from socioeconomically disadvantaged backgrounds may have higher rates of Long COVID. This reflects a broader pattern of health disparities that is also observed for other diseases.

Various factors contribute to these differences. One of the key factors is differential exposure to the virus, which is often influenced by working conditions, as people from lower socio-economic backgrounds are more likely to work in jobs that they cannot do remotely and that require regular contact with other people. In addition, ethnic minorities and socio-economically disadvantaged groups often have poorer access to health services, which can delay early diagnosis and treatment, thus increasing the risk of long-term consequences such as Long COVID.

Other social determinants of health, such as housing conditions, food security and general living conditions, also play a role. Cramped living conditions and reduced access to healthcare can increase transmission rates and at the same time hinder recovery from the disease.

Ongoing research on Long COVID is crucial to better understand these complex interactions and to develop effective strategies targeted to the most affected groups. Future studies and more comprehensive data sets will enable the development of tailored interventions that include not only medical but also social support measures to reduce health inequalities and improve resilience to Long COVID across all populations.

B. Findings on the possible causes of Long COVID

The scientific community is intensively investigating the possible causes of Long COVID in order to develop effective treatments and management strategies. There are several hypotheses that attempt to explain the persistent symptoms and the variety of experiences of those affected. Two key areas that are particularly emphasised in research are viral persistence and immune responses.

I. Viral persistence

The hypothesis that persistent virus particles or virus fragments may play a role in Long COVID is one of the key ideas currently being explored to explain the diverse and persistent symptoms that some people experience after infection with SARS-CoV-2. This hypothesis is based on observations and study results that suggest that in some cases, the virus or parts of it may remain detectable in certain body tissues for longer than expected.

The possibility that RNA fragments or even intact SARS-CoV-2 viruses remain in various organs after the acute infection phase opens up important perspectives for understanding Long COVID. Such findings are based on detailed studies showing the presence of viral material in tissues long after the initial recovery. These observations have led to several theories on how persistent viral material could cause long-term health problems.

a. Viral persistence and its possible effects

- Survival of the virus in cell reservoirs: The hypothesis that the virus survives in certain cell types is based on the discovery that SARS-CoV-2 is able to colonise cells that are not normally renewed quickly. For example, cells in the gastrointestinal tract or neurons in the central nervous system could serve as long-term reservoirs for the virus. As these cells have a longer lifespan and are not regularly renewed, the virus could remain in them and cause a chronic, albeit weak, immune response. This persistent immune activity can lead to inflammation, which manifests itself through various symptoms such as fatigue, brain fog and digestive problems.
- Chronic inflammatory reactions: Even if the virus is no longer infectious, the remaining viral RNA fragments can continue to act as antigens that activate the immune system. This ongoing immune response can lead to chronic inflammation that affects various organ systems, contributing to the multiple symptoms of Long COVID.
- Autoimmune reactions: A further theory concerns the possibility of an autoimmune reaction. If viral proteins or RNA fragments persist, the

immune system could mistakenly recognise the body's own cells, which have similar molecular patterns, as foreign and attack them. This auto-immune reaction could cause a wide range of symptoms, depending on which cells or tissues are affected.

b. Research directions and challenges

Research into these theories presents scientists with several challenges. Firstly, it must be confirmed in which tissues and to what extent viral fragments or intact viruses can remain. This requires advanced virological and histopathological techniques. Secondly, the causal relationship between the presence of these fragments and the symptoms of long COVID must be clarified, which requires the development of models that can mimic these chronic conditions.

Finally, the development of effective treatments for long COVID requires a better understanding of the underlying mechanisms. These include potential antiviral therapies that specifically eliminate remaining viral reservoirs, as well as immunomodulators that can correct the excessive or misdirected immune response triggered by the virus.

Ongoing research on these aspects will be crucial to better understand and treat the long-term effects of SARS-CoV-2. Given the global scope of the pandemic and the large number of people potentially suffering from Long

COVID, it is crucial to decipher the mechanisms behind these persistent symptoms.

c. Effects on organs

The persistence of the virus in organs such as the gut or brain could explain many of the reported long-term symptoms of Long COVID. For example, persistent viral presence in the gut could lead to long-term changes in the gut flora and associated digestive problems. In the brain, persistent viral activity could lead to neurological and cognitive impairments, including the often-reported "brain fog".

Research in this area is still in its infancy and the exact mechanisms by which persistent viral particles or fragments cause long COVID symptoms are not yet fully understood. Extensive studies are needed to:

- To determine the prevalence and localisation of persistent virus particles in different tissues.
- To clarify the mechanisms by which these particles could trigger persistent symptoms.
- To understand the relationship between virus persistence and the severity and duration of long COVID symptoms.

Further research into this hypothesis could not only provide important insights into the pathology of Long COVID, but also lead to more targeted treatment strategies aimed at eliminating or controlling the sources of persistent viral activity in the body. This could ultimately help to significantly improve the quality of life of Long COVID patients.

II. Immune reactions

The theory that Long COVID is caused by a dysregulation or overreaction of the immune system is another key assumption in current research on the long-term effects of COVID-19. This immune response, which remains active after the acute infection has subsided, may play a central role in the development of the persistent and often debilitating symptoms of Long COVID.

a. Immune dysregulation for long COVID

The immune system is designed to fight infections and protect the body from damage. Normally, it should return to a dormant state after successfully fighting a pathogen such as SARS-CoV-2. In some Long COVID patients, however, this return process appears to be disrupted. The immune system remains in an activated state, possibly due to residual viral antigens or a misguided perception of the body's own material as foreign.

b. Chronic inflammatory reactions

One of the main consequences of such persistent immune activity is chronic inflammation. Long COVID patients have been found to have elevated levels of cytokines, which are important mediators in the immune response. These cytokines, including interleukins and tumour necrosis factors, are essential in controlling inflammatory processes. Their increased presence can lead to a number of symptoms, such as

- Fatigue: One of the most common symptoms of long COVID, often associated with general immune dysfunction.
- Joint pain: Inflammation in the joints or surrounding tissue can lead to persistent pain.
- Neurological impairments: The cognitive impairments known as "brain fog" can be exacerbated by neuroinflammatory processes that impair brain function.

c. Autoimmunity

The theory that autoimmune reactions play a significant role in Long COVID is based on the concept of molecular mimicry and the release of hidden antigens. These mechanisms can lead to the immune system remaining active or even attacking the body's own cells after a SARS-CoV-2 infection.

aa. Molecular mimicry

Molecular mimicry occurs when parts of the virus structures are similar to those of certain endogenous proteins.

The immune system, which is trained to recognise and fight the virus, can then mistakenly attack the body's own cells that resemble these viral proteins. This mechanism could lead to a variety of symptoms, depending on which cells or tissues are affected. For example, neurological symptoms could arise if neuronal proteins become the target of such a misdirected immune response.

bb. Release of antigens

During the acute infection phase, normally hidden antigens can be released due to cell damage. These exposed antigens can then be recognised as foreign by the immune system, leading to a further autoimmune reaction. This reaction can cause chronic inflammation and associated symptoms such as joint pain and persistent fatigue.

cc. Studies on immune reactions and cytokine patterns

Research is focussed on deciphering the specific patterns of immune response, including cytokine production, in long COVID patients. Cytokines, small proteins released by immune cells, play a central role in controlling the inflammatory response. By understanding these patterns, researchers can better identify which immune responses are dysregulated in Long COVID.

dd. Development of biomarkers and immuno-modulators

A key area of current research is the identification of biomarkers that enable early diagnosis of Long COVID. These biomarkers could also help predict the severity of the disease and select the most appropriate therapeutic approaches. At the same time, therapies are being researched that can specifically modify or suppress the dysregulated immune responses. Immunomodulators could help to dampen the excessive immune response and at the same time maintain the necessary immune defence against other pathogens.

The challenge is to develop therapies that are specific enough to modulate the dysregulated immune response in long COVID without weakening the overall function of the immune system. Future research approaches must also consider the long-term effects of these therapies and ensure that they do not lead to undesirable side effects.

A deeper understanding of the role of the immune system in the pathogenesis of Long COVID is crucial to develop effective and safe treatments that improve the quality of life of those affected without compromising their overall immunity. This requires close collaboration between researchers, clinicians and patients to develop customised and effective intervention strategies.

III Other causes of long COVID

There are also other factors and mechanisms that are being researched as possible causes of long COVID.

a. Vascular damage

COVID-19 can have significant effects on the vascular system and trigger blood clotting disorders.

One of the most remarkable effects of COVID-19 is its tendency to increase blood clotting. The virus can trigger hypercoagulability, a condition in which the blood clots more easily. This can lead to thromboses and embolisms, in which blood clots form in the blood vessels and block them. Such events are particularly dangerous if they occur in the large or small vessels of the brain, heart or lungs.

b. Microvascular damage

COVID-19 is also associated with damage to the small vessels, the so-called microvessels. These small vessels are responsible for the blood supply to organs and tissues. Damage to them can lead to a reduced blood supply, which can impair the function of the affected organs. The impairment of microcirculation can lead to a lack of oxygen in various tissues, causing a range of symptoms such as fatigue and weakness.

c. Contribution to long COVID symptoms

- Fatigue: One of the most common complaints of Long COVID is fatigue, which can be exacerbated by a variety of factors, including vascular problems. Poor blood circulation due to microvascular damage can affect the supply of oxygen and nutrients to cells, leading to persistent fatigue and lack of energy.
- Brain fog: Neurological symptoms such as difficulty concentrating and memory problems can be caused in part by microvascular damage in the brain. Impaired blood circulation in the brain can lead to a sub-optimal supply of oxygen and other important nutrients to neuronal cells, which impairs cognitive function.

d. Research approaches and therapeutic options

Research is increasingly focussed on better understanding and treating these vascular complications. Therapeutic approaches could include drugs that modulate blood coagulation, improve microcirculation or strengthen endothelial function. Such treatments aim to normalise blood flow dynamics and thereby reduce the symptoms caused by vascular damage.

Thus, it can be said that the vascular effects of COVID-19 may be a central factor in the development of Long COVID. Continued research and development of targeted therapies are crucial to understand the mechanisms leading to these long-term health problems and to provide effective treatments that improve the quality of life of those affected.

IV. Neurological effects

The impact of COVID-19 on the nervous system is another key area of research that is receiving increasing attention as evidence is accumulating of both direct and indirect effects of the virus on the brain and wider nervous system. These effects may have long-term consequences and contribute to the neurological symptoms that some patients experience as part of Long COVID.

a. Direct invasion of brain cells by the virus

There is increasing evidence that SARS-CoV-2 has the ability to enter brain cells directly. This could occur by utilising the ACE2 receptor, which is expressed not only in the respiratory tract but also on neurons and other cells in the central nervous system. Direct invasion of the virus can lead to viral encephalitis, an inflammation of the brain that can cause neuronal damage. Such damage can manifest itself in a variety of symptoms, including cognitive deficits, memory problems, confusion and, in severe cases, neurological impairments such as seizures.

b. Indirect effects due to inflammatory processes

In addition to direct viral invasion, there are also indirect mechanisms by which COVID-19 can affect the nervous system:

- Cytokine storm: A severe acute reaction to SARS-CoV-2 can lead to an excess of inflammatory cytokines, a phenomenon known as a cytokine storm. This excessive immune response can also affect the brain, leading to inflammation that can affect the blood-brain barrier and disrupt cognitive function.
- Autoimmune reactions: As mentioned above, COVID-19 can trigger autoimmune-like reactions in which the immune system mistakenly attacks the body's own neuronal cells. These reactions can lead to persistent inflammation in the brain, causing neurological and cognitive symptoms.
- Vascular damage: COVID-19 affects the vascular system, which can lead to impaired blood flow to the brain. This can lead to ischaemic attacks and microvascular damage in the brain, which impairs cognitive function.

V. Endothelial dysfunction

Endothelial damage caused by COVID-19 is another significant aspect of the pathophysiological impact of the virus on the human body and plays a crucial role in many of the persistent symptoms that occur in the context of Long COVID. The endothelium, the innermost layer of blood vessels, plays a central role in regulating various vascular functions, including blood clotting, inflammatory responses and blood flow.

a. Role of the endothelium

The endothelium not only acts as a passive barrier, but is also actively involved in the regulation of blood flow and vessel width. It produces and secretes a variety of substances that can have a vasodilating (vasodilating) and vasoconstrictor (vasoconstricting) effect. It also plays a decisive role in immunomodulation and the body's inflammatory response.

b. Mechanisms of endothelial damage caused by COVID-19

- Direct viral invasion: SARS-CoV-2 can bind directly to ACE2 receptors on endothelial cells, which are mainly present in arteries, veins and capillaries throughout the body. This direct binding can lead to invasion and infection of endothelial cells, which can disrupt their function and lead to their death.

- Inflammatory reactions: Infection of the endothelial cells can trigger a strong local inflammatory reaction. The immune system responds to the infected cells by releasing cytokines and other inflammatory mediators, leading to further damage to the endothelium. This inflammation can contribute to endothelial dysfunction, in which the balance between vasodilating and vasoconstrictor substances is disturbed.

- Oxidative stress: The oxidative stress induced by the virus and the inflammatory response can cause additional damage to the endothelium. Oxidative stress occurs when reactive oxygen species (ROS) are produced at a level that exceeds the body's antioxidant capacity. This leads to oxidation and damage to the endothelial cells.

c. Clinical effects of endothelial damage

Damage to the endothelium has far-reaching effects on vascular health and can lead to a range of symptoms and diseases:

- High blood pressure: Due to the dysfunction of the endothelium, less nitric oxide, an important vasodilator, is produced. This can lead to general vasoconstriction and increased blood pressure levels.

- Breathing problems: Endothelial damage in the pulmonary vessels can lead to impaired blood circulation in the lungs, which impairs gas exchange and leads to breathing difficulties.
- Thrombosis and embolism: Impaired endothelial function increases the risk of thrombosis, as the endothelium normally has anticoagulant properties. Damage can therefore lead to an increased tendency to clot, which increases the risk of venous thromboembolism.

d. Research and treatment

Research is focussed on the development of therapies aimed at supporting and restoring endothelial health. These may include antioxidants, anti-inflammatory drugs, drugs to lower blood pressure and substances to improve endothelial function. The use of statins and other vascular protective agents is also being investigated to counteract the endothelial damage caused by COVID-19.

The importance of endothelial integrity to overall health highlights the importance of a comprehensive understanding of the vascular impact of COVID-19 for the treatment and management of Long COVID.

VI Genetic, physiological and environmental risk factors

Long COVID is a complex syndrome whose risk factors are a combination of genetic, physiological and environmental influences. Research into these risk factors is still ongoing, but findings to date provide important insights that can help to identify risk groups and develop prevention strategies.

a. Genetic factors

Genetic predispositions play a significant role in an individual's response to SARS-CoV-2 infection and can significantly influence how a person recovers from the infection and whether they develop Long COVID. Various studies have shown that certain genetic markers, particularly those associated with the immune system, are more common in individuals who develop Long COVID. These genetic differences may make certain people more susceptible to prolonged inflammatory responses and other long-term symptoms.

aa. Genetic markers and their effects

- Cytokine production: Cytokines are key proteins that play an essential role in the immune response by regulating inflammatory processes. Variations in genes responsible for the production and regulation of cytokines can influence the way the body responds to infection. Polymorphisms in these genes could lead to excessive or prolonged production of proinflammatory cytokines, which increases the risk of chronic inflammation and thus Long COVID. Examples of such cytokines are interleukin-6 (IL-6), tumour necrosis factor-alpha (TNF-α) and interferons.

- Immune cell receptors: Genes that code for receptors on immune cells play a central role in recognising pathogens and initiating the immune response. Variations in these genes can influence the efficiency and specificity of the immune response. For example, polymorphisms in the gene for the ACE2 receptor, which serves as the entry point for SARS-CoV-2 into cells, could influence susceptibility to infection and the severity of the disease.

bb. Specific genetic variations

- HLA genes: The human leukocyte antigen (HLA) system is crucial for antigen presentation and activation of the adaptive immune system. Certain HLA genotypes have been associated with increased susceptibility to severe courses of COVID-19 and possibly also Long COVID. HLA variations could influence the ability of the immune system to effectively recognise and fight viral antigens.

- IFITM3: The gene for interferon-induced trans-membrane protein 3 (IFITM3) is known to play a role in the defence against viral infections. Variations in this gene have been associated with varying degrees of severity of viral respiratory infections, including COVID-19. Reduced function of IFITM3 could lead to ineffective viral defence and an increased inflammatory response.

Identifying and analysing these genetic markers can provide important insights into why some people are more susceptible to Long COVID than others. This opens up new possibilities for personalised medicine approaches, where treatments and prevention strategies can be developed based on a patient's individual genetic profile.

Some of the current research directions include:

- Genome-wide association studies (GWAS): These studies examine large populations to identify genetic variants associated with the risk of Long COVID.
- Functional genomics: These approaches aim to understand the functional effects of identified genetic variants and clarify how they influence the immune response and inflammatory processes.

- Biomarker development: The identification of genetic markers associated with an increased risk of Long COVID could lead to the development of biomarkers that allow early identification of individuals at risk.

cc. Clinical relevance

Understanding the genetic predispositions to Long COVID is crucial for the development of preventive measures and therapeutic interventions. Individuals at higher risk for Long COVID due to their genetic makeup could benefit from targeted monitoring and treatment strategies specifically aimed at minimising the long-term impact of the disease. The clinical relevance of understanding genetic predispositions to Long COVID spans several key areas of medical research and practice.

dd. Early detection and risk assessment

By understanding genetic factors that influence the risk of Long COVID, doctors can identify potentially at-risk individuals at an early stage. This enables proactive risk assessment and the implementation of preventive measures. For example, people with certain genetic markers that indicate a higher risk of severe Long COVID symptoms could be monitored more closely or included in special follow-up programmes at an early stage.

ee. Personalised therapy approaches

Genetic insights can lead to the development of customised therapies that are tailored to the specific pathophysiologies prevalent in individual patients. For example, individuals with genetic variants that lead to excessive cytokine production could benefit from treatments that specifically target inflammation, while others who may be more prone to endothelial dysfunction could receive drugs that specifically support vascular health.

ff. Improvement of clinical results

Early identification and personalised treatment can reduce the likelihood of long-term health problems and improve the quality of life of those affected. This is particularly important in a disease like Long COVID, where symptoms can vary greatly and many different organ systems can be affected.

gg. Development of preventive medicine

Knowledge of genetic predispositions can also be utilised in preventive strategies. For example, the development of vaccine strategies tailored to genetic risk factors could be an effective way to prevent severe courses and the development of Long COVID. Similarly, adapting lifestyle recommendations to genetic risk profiles could help to strengthen resilience to the disease.

At the population level, understanding genetic factors can help to allocate resources more efficiently by prioritising support for at-risk groups. This could lead to a more targeted public health strategy aimed at minimising the overall impact of COVID-19.

Overall, understanding the genetic predispositions to Long COVID provides a fundamental basis for the development of innovative, effective and personalised medical interventions that have the potential to significantly improve the prevention, treatment and long-term management of COVID-19.

b. Physiological factors

The physiological factors that influence the risk and severity of Long COVID are diverse and complex. In addition to age, gender and pre-existing medical conditions, genetic predispositions, immune responses and lifestyle factors also play a significant role in the development and progression of Long COVID.

aa. Age and gender

As we age, immune function typically declines, a phenomenon known as immune senescence. This can reduce the body's ability to effectively respond to and recover from viral infections. Older people often have a reduced T-cell response, which makes viral infections more difficult to control and increases the risk of long-lasting symptoms.

Women experience Long COVID more frequently than men. This could be due to differences in the immune response, some of which are hormonal. Oestrogens can influence the immune system by modulating both the level of immune response and tolerance to one's own tissue. Women also have a higher incidence of autoimmune diseases, indicating a more pronounced immune response that could play a role in COVID-19.

bb. Pre-existing medical conditions

Conditions such as diabetes, cardiovascular disease and obesity are not only risk factors for a severe course of COVID-19, but also for the development of Long COVID. These diseases are often associated with chronic low-grade inflammation, which can weaken the immune system and increase susceptibility to prolonged inflammatory responses.

Chronic diseases can impair the functioning of the immune system, which can lead to an inefficient or excessive response to SARS-CoV-2. Such a dysregulated immune response may play a key role in the development of persistent symptoms of Long COVID.

cc. Severity of the original COVID-19 disease

A severe course of the initial COVID-19 disease is a predictor of Long COVID. Severe cases can lead to a stronger and longer-lasting activation of the immune system, which increases the likelihood of dysregulation and the resulting long-term consequences.

Severe COVID-19 cases are often characterised by pronounced inflammatory responses, including elevated levels of cytokines and other inflammatory markers. This strong inflammatory response can damage tissue and complicate recovery, leading to a variety of long-term symptoms.

dd. Additional factors

Smoking, low physical activity and an unhealthy diet can increase the risk of severe COVID-19 and the development of Long COVID. These factors can increase the general tendency to inflammation and weaken the immune response.

Access to healthcare, socioeconomic status and working conditions can also influence how a person responds to and recovers from the infection. Individuals in lower socioeconomic strata or those working in high-risk environments may be at higher risk of developing severe forms of COVID-19 and Long COVID.

The combination of these factors can influence the risk and severity of Long COVID and explains the wide variability in experiences

c. Environmental factors

Environmental factors also contribute significantly to the development and progression of Long COVID. These factors influence not only the likelihood of infection with SARS-CoV-2, but also the way the body reacts to the virus and recovers from the acute phase of the disease.

aa. Exposure and working environment

People in professions with a high viral load, especially in the healthcare sector, are not only at increased risk of initial infection, but also of more frequent and potentially more intense exposure to the virus. This can lead to a more severe acute infection, which in turn increases the risk of developing Long COVID. Constant exposure to the virus can overwhelm the body's immune response and lead to chronic inflammation, a key factor in Long COVID.

Inadequate protective measures and equipment can increase the risk of infection. Work environments that implement appropriate protective measures can reduce the risk of serious illness and thus also the risk of Long COVID.

bb. Socio-economic factors

Living close together in densely populated areas can increase transmission rates and often makes isolation more difficult in the event of illness, which in turn can affect recovery. Such conditions are more common in lower socio-economic strata.

Limited access to healthcare and treatment can prevent early detection and effective treatment of COVID-19, increasing the likelihood of complications and the development of Long COVID. People in lower socioeconomic groups often have less access to healthcare services, which increases their vulnerability.

Financial insecurity, job worries and the general stress that comes with lower socio-economic conditions can weaken the immune system and delay recovery.

cc. Lifestyle factors

A nutrient-rich diet strengthens the immune system and supports the body's own healing processes, while an unbalanced diet can increase the risk of chronic inflammation and delay recovery.

Regular exercise can improve overall health, strengthen the immune system and reduce the risk of chronic diseases that could increase the risk of severe COVID-19 and Long COVID.

Smoking damages the lungs and other body tissues, weakens the immune defence and increases the risk of severe courses of COVID-19 and the development of long COVID symptoms.

Attention to these environmental and lifestyle-related factors is crucial for the prevention and management of Long COVID. Targeted public health strategies and individual preventive measures can minimise risks and improve overall resilience to Long COVID.

VII The role of the initial COVID-19 disease severity

The severity of the original COVID-19 disease plays a significant role in the risk assessment for the development of Long COVID, although the picture is complex and not entirely predictable. Some key correlations and observations help to understand this influence:

The relationship between the severity of initial COVID-19 infection and the risk of developing Long COVID is a critical area of research that has important implications for the management and treatment of COVID-19. Several studies suggest that individuals with more severe initial symptoms and those who were hospitalised during their infection have an increased risk of Long COVID. The reasons for this are complex and related to the type of immune response and the body systems affected.

a. Severity of the infection and inflammatory reaction

A more severe infection typically triggers a stronger inflammatory reaction. This response is characterised by high levels of cytokines and other inflammatory markers, often referred to in medical jargon as a "cytokine storm".

Although this strong immune response may be necessary in the short term to fight the virus, it can also lead to tissue damage and thus increase the likelihood of long-term complications.

In severe cases, the body's immune response can go into overdrive, leading to an autoimmune reaction in which the immune system mistakenly attacks healthy cells and tissues. This can cause long-lasting or even permanent damage to various organs, increasing the likelihood of Long COVID symptoms.

b. Affected systems during the acute infection

Patients with severe courses of COVID-19 often experience cardiovascular symptoms such as arrhythmias, increased blood pressure and damage to the heart vessels caused by inflammation. These problems can persist in the long term and lead to chronic cardiovascular problems, which are often seen in Long COVID.

Severe cases of COVID-19 can also cause lung problems, including pneumonia and acute respiratory distress syndrome (ARDS). These conditions can lead to long-term damage to lung tissue, resulting in persistent breathing problems that are often a core symptom of Long COVID.

c. Implications for treatment and prevention

Understanding the link between the severity of the initial infection and Long COVID has important clinical implications.

Early and aggressive treatment of COVID-19, especially in people who show signs of severe disease, could help reduce the severity of the inflammatory response and reduce the risk of Long COVID.

Individuals who have survived a severe COVID-19 infection should be closely monitored for signs of Long COVID and, if necessary, be supported in specialised rehabilitation programmes aimed at restoring the function of various affected systems.

Based on the severity of the disease and the body systems affected, individualised treatment plans can be developed that aim to treat specific symptoms and damage that can occur with Long COVID.

Thus, understanding the link between the severity of the initial COVID-19 infection and the risk of Long COVID is crucial to develop effective treatment and prevention strategies aimed at minimising the long-term health impact of this global pandemic.

d. Immune system and inflammatory response

A more severe initial infection with COVID-19 can often trigger a more intense and sometimes misdirected immune response, leading to long-lasting inflammation. These inflammatory processes are not limited to the acute phase of the disease, but often persist and affect the body's recovery over a longer period of time. The persistent inflammation can damage various body systems, complicating the healing process and increasing the risk of developing Long COVID. In cases where the immune system remains overactive and turns against its own body, this can lead to a variety of symptoms that last for months and significantly impact quality of life. The ongoing burden of inflammation can also disrupt the function of key organs and lead to a wide range of health problems that further delay full recovery.

e. Existing organ damage

Severe cases of COVID-19 are often associated with direct damage to vital organs, which can lead to long-term functional impairment. This damage is often profound and primarily affects the lungs, heart and kidneys, resulting in a variety of persistent symptoms that characterise Long COVID. Affected individuals experience

reduced lung function, which manifests itself in shortness of breath that can occur even with little physical exertion. In addition, cardiovascular problems caused by
the infection, such as cardiac arrhythmias, can persist
even after the acute virus has subsided. This can lead to
a feeling of constant exhaustion, as the heart may no
longer be able to supply the body with blood and oxygen efficiently. The kidneys, which play a crucial role in
filtering waste and regulating fluids, may also be affected, further reducing overall health and well-being.

These long-term organ impairments can lead to a vicious
circle of chronic fatigue, reduced physical performance
and reduced quality of life. In addition, persistent organ
inflammation as a result of the initial damage can delay
the healing process and require long-term medical monitoring and treatment. Such long-term effects often require a multidisciplinary approach to medical care, including specialised rehabilitation programmes aimed at
improving the function of the affected organs and minimising the progression of further damage.

f. Psychological effects

More severe cases of COVID-19 can result in psychological and cognitive effects that go far beyond the direct
physical effects of the disease. The psychological stress
associated with a severe illness, a possible long hospitalisation and the associated experiences of isolation, fear

and uncertainty can lead to a variety of long-term psychological and cognitive problems.

The psychological burden often begins with the stress and anxiety associated with the diagnosis and the uncertainty about the outcome of the disease. Patients who require intensive medical interventions, such as treatment in intensive care or mechanical ventilation, can go through particularly traumatic experiences. Such experiences can increase the risk of developing post-traumatic stress disorder (PTSD), anxiety disorders and depression.

This emotional and psychological stress can also impair cognitive function. Many people who survive severe COVID-19 disease report persistent "brain fog", which manifests itself in concentration problems, memory difficulties and slowed thought processes. These symptoms can result from the combination of the direct effects of the virus on the brain and the stress-related effects on mental health.

In addition, the psychological effects of the illness can impair physical recovery. For example, stress and anxiety can exacerbate inflammatory processes in the body or weaken the immune system, further delaying recovery from the physical illness. This creates a cycle in which psychological problems affect physical health, which in turn leads to further psychological stress.

The treatment and support of patients with severe COVID-19 should therefore include not only the physical, but also the psychological and cognitive aspects. Comprehensive care could include psychotherapeutic support, cognitive rehabilitation and, if necessary, psychiatric medication. The involvement of support networks, the provision of information resources and the promotion of stress management strategies are also crucial to support recovery and improve the wellbeing of those affected.

Although a severe course of COVID-19 disease increases the risk of Long COVID, it is important to emphasise that individuals with initially mild symptoms can also develop Long COVID. This suggests that additional factors such as genetic predispositions, immune response and possibly unknown viral characteristics also play a role. Research into these relationships remains dynamic and will continue to be pursued by the scientific community to develop a more complete understanding of the pathophysiology of Long COVID.

VIII. Links between vaccination and Long COVID

Vaccination against COVID-19 plays a central role in the prevention of severe disease progression and also has implications for the prevention and management of Long COVID. Studies and clinical data suggest that vaccines can not only reduce the risk of severe acute COVID-19 infection, but also reduce the risk of developing Long COVID.

a. Reducing the risk of serious illness

COVID-19 vaccines play a critical role in combating the pandemic by reducing the risk of infection with the SARS-CoV-2 virus or, if infection occurs, significantly reducing the severity of the disease. This ability of the vaccines to reduce the severity and duration of the disease has a direct impact on healthcare systems and the individual health of the population.

By reducing severe symptoms, vaccinations often prevent people from needing intensive care treatments. Intensive care treatments, such as mechanical ventilation or long-term hospitalisation, are often associated with an increased risk of complications, including the development of Long COVID. Post-viral fatigue and other long-lasting symptoms of Long COVID can be attributed in part to the extreme stress that severe COVID-19 cases place on the immune system.

Vaccinated individuals who contract COVID-19 usually experience milder symptoms that are less likely to lead to serious or long-term health problems. This is because vaccination primes the immune system to respond more effectively and quickly, reducing viral load and the associated immune response that can lead to tissue damage. A lower viral load means less chance of extensive inflammatory processes and therefore less chance of long-term damage that could lead to Long COVID.

In addition, a shorter duration of illness can limit the amount of time the immune system is heavily challenged. A quick recovery reduces the risk of the immune system entering a state of chronic activity or dysregulation, which in turn reduces the risk of long-lasting symptoms.

Overall, vaccines provide critical protection that goes beyond the immediate prevention of the disease and offers important long-term health benefits. They help to reduce the burden on healthcare systems, minimise exposure to the virus in the population and help to reduce the impact of the pandemic on public life and the economy. By reducing the risk of severe illness and consequently of Long COVID, vaccinations are an essential tool in the fight against the COVID-19 pandemic.

b. Reducing the risk of long COVID

The effectiveness of COVID-19 vaccines in reducing the risk of Long COVID is based on several important mechanisms that are supported by clinical trials and research data. The vaccines play a crucial role by not only reducing the overall likelihood of infection, but also mitigating the severity of symptoms if infection does occur. These effects contribute significantly to reducing the long-term health consequences of COVID-19, including Long COVID.

aa. Reduced infection rate

Vaccinations increase immunity to the SARS-CoV-2 virus and thus reduce the likelihood of infection. A lower infection rate within the vaccinated population automatically means a lower number of people who could potentially develop Long COVID. This direct reduction in the incidence of infection is the first protective step offered by vaccination.

bb. Milder course of the disease

Even if vaccinated people are infected with COVID-19, data shows that the course of the disease tends to be milder. A milder course often means a lower viral load and less stress on the immune system. This in turn minimises the likelihood of an excessive immune response, which can lead to the persistent symptoms of Long COVID. Reducing inflammatory responses and avoiding severe symptoms such as high fever, persistent cough and respiratory problems can reduce the risk of long-term damage to organs such as the lungs, heart and kidneys.

cc. Shorter duration of illness

Vaccination can also shorten the duration of the disease, which reduces the window of acute inflammation and thus the duration of stress on the body. A shorter illness phase gives the body the opportunity to switch to recovery mode more quickly, increasing the chance of a full and uncomplicated recovery and minimising the development of long COVID symptoms.

dd. Public health and herd immunity

In addition to individual protection, vaccinations also contribute to herd immunity. The more people are vaccinated, the lower the spread of the virus in the community. This reduces the number of new COVID-19 cases and thus indirectly the number of potential long COVID cases. Lower circulation of the virus also limits the possibility of reinfection, which can lead to repeated acute phases and possibly to Long COVID.

Vaccination therefore remains a central element of the strategy against COVID-19 and its long-term consequences. It is important that this message continues to be communicated and supported in order to increase vaccination rates and protect the health of the population.

Thus, it can be said that vaccination is an important tool not only to mitigate the acute effects of COVID-19, but also to reduce the incidence and severity of Long COVID, reducing the health burden on individuals and health systems worldwide.

C. Common symptoms of Long COVID and their effects

Long COVID can include a variety of physical symptoms that have a profound impact on the daily functioning and quality of life of those affected. The most common and debilitating symptoms include fatigue, breathing difficulties and cardiovascular problems.

I. Exhaustion (fatigue)

a. Description of the symptoms

The fatigue that many people with Long COVID experience is one of the most commonly reported and most difficult symptoms to treat. This form of fatigue differs significantly from normal fatigue as it is profound and not necessarily related to previous physical or mental exertion. The intensity and persistence of this fatigue can have a significant impact on the quality of life of those affected.

b. Characteristics of exhaustion with long COVID

Fatigue in Long COVID is often disproportionate to the activity or exertion performed and is not relieved by rest or sleep. This phenomenon is also known as postexertional malaise (PEM), a condition in which symptoms worsen after physical or mental exertion and can persist for a disproportionate amount of time. Sufferers often describe their energy as being 'switched off' and that even minimal efforts such as showering, shopping or light household chores become overwhelming challenges.

c. Possible causes

The exact mechanisms behind Long COVID fatigue are not yet fully understood, but researchers suspect a combination of immunological, neurological and metabolic factors:

- Persistent or misdirected immune system activity after recovery from the acute COVID-19 phase could lead to chronic inflammatory processes that cause fatigue.
- The virus could cause structural and functional changes in the brain that lead to exhaustion. This could include impairment of the mitochondria, which are important for energy production in the cells.
- The infection and associated stress could affect the endocrine system, leading to changes in the production of hormones that control energy regulation and mood.
- Some sufferers experience dysregulation of the autonomic nervous system, which regulates the

heartbeat, blood pressure and breathing, among other things, which can lead to a loss of energy.

d. Effects on daily life

The fatigue that many people with Long COVID experience is a significant hindrance to their everyday lives. This profound fatigue not only reduces physical and mental energy for daily tasks, but also has a serious impact on work and social life. Many sufferers find themselves in a state where even basic activities such as preparing meals, shopping or maintaining personal hygiene can be overwhelming. The difficulty in managing day-to-day routines and responsibilities can lead to a significant reduction in the ability to work, with some people having to severely restrict or give up their professional activities altogether.

These restrictions in working life often have far-reaching financial and emotional consequences, as the loss of the ability to work can affect self-esteem and cause financial insecurity. In addition, prolonged exhaustion can lead to those affected becoming socially isolated because they no longer have the energy to socialise. This withdrawal can lead to loneliness and depression, which further affects mental health.

Social isolation and the loss of participation in professional and private life can create a vicious cycle in which isolation exacerbates exhaustion and the associated symptoms, which in turn intensifies isolation. The resulting psychosocial stress often exacerbates the physical symptoms of Long COVID, making recovery even more difficult. To meet these challenges, those affected require comprehensive support that includes medical, psychological and social interventions. Approaches such as targeted therapies, adapting the workplace and support in the social environment can help to improve the quality of life of those affected and promote their participation in social and professional life.

e. Treatment approaches

There is currently no specific treatment for Long COVID fatigue, but there are approaches that can help manage the symptoms:

aa. Pacing techniques

Pacing techniques are an essential approach for people struggling with chronic fatigue or long COVID. These methods help those affected to carefully distribute their available energy throughout the day or even the week in order to conserve physical and mental resources and avoid overexertion. The basic principle of pacing is to consciously plan and carry out activities in such a way that they do not overtax the individual's energy balance and leave enough time for recovery.

Pacing teaches people to prioritise their daily activities, break them down into smaller, manageable chunks and take regular breaks before exhaustion sets in. This may mean performing some tasks more slowly or with more frequent rest periods to ensure constant energy availability. It's about finding a balance between activity and rest that doesn't exacerbate symptoms and allows the body not to overexert itself.

For example, someone who could easily work for several hours without a break before becoming ill may now realise that a short break is needed after every 30 minutes to maintain energy levels and avoid exhaustion. It could also be helpful to spread heavy physical activities over days when there are fewer other demands in order to avoid overworking.

The use of pacing techniques often requires an initial adjustment phase and a change in thinking, as many people are used to organising their activities around external demands rather than their own energy levels. It is therefore also important to inform the personal and professional environment about these adjustments so that there is understanding and support for the new limits and needs.

In addition, keeping a diary in which activities and energy levels are recorded can help to recognise patterns and better understand one's own limits. This enables those affected to plan their days more effectively and use their energy wisely.

bb. Rehabilitative therapies

Rehabilitative therapies such as physiotherapy and occupational therapy play a crucial role in the treatment and recovery of individuals suffering from Long COVID. These therapeutic approaches aim to gradually restore physical capacity and functionality, which is particularly important as Long COVID can lead to a wide range of physical and cognitive impairments.

Physiotherapy focuses on gradually increasing movement and physical activity to improve muscle strength, endurance and flexibility. For people suffering from Long COVID, this type of therapy can be crucial, as many suffer from muscular weakness, fatigue and pain that significantly limits their mobility and quality of life. Physiotherapists develop personalised exercise plans that aim to increase physical resilience without overexertion. These plans are often designed to include cardiovascular health and respiratory aspects, as COVID-19 can affect the respiratory system in particular.

Occupational therapy focusses on restoring the ability to perform everyday tasks and supporting people to gain independence in their daily lives. Occupational therapists work closely with those affected to identify specific challenges in their daily lives and develop practical solutions to help them cope with the effects of Long COVID. This may include adapting the home environment to make it safer and more accessible, or learning new strategies to manage energy and reduce the stress of everyday activities. Occupational therapists also offer cognitive rehabilitation for those suffering from memory problems or 'brain fog', which is also a common symptom of Long COVID.

Both forms of therapy often also focus on education and self-management strategies to teach patients how to best manage their symptoms. This includes techniques to reduce stress, manage pain and avoid activities that could potentially exacerbate symptoms.

The integrative approach of these rehabilitation methods is crucial, as it takes into account not only the physical, but also the psychological and social aspects of recovery. This comprehensive approach helps those affected to regain their independence and quality of life step by step, which is essential for overcoming the complex challenges that Long COVID entails.

cc. Psychological support

Psychological support also plays a crucial role in coping with the far-reaching emotional and psychological effects of Long COVID. Many sufferers experience not only physical symptoms, but also significant psychological distress caused by the ongoing illness and associated life changes. Counselling and cognitive behavioural therapy (CBT) are two forms of psychological support that have been shown to be particularly effective in helping individuals overcome these challenges.

Counselling provides a safe space in which people can express their fears, worries and the psychological stress caused by Long COVID. These conversations help those affected to process their experiences and develop strategies to deal with the daily challenges. An important aspect of counselling is validating the patient's experience, which is particularly important as Long COVID often has invisible symptoms that are difficult for outsiders to understand. This can increase feelings of isolation and not being understood. Counsellors can also help plan lifestyle and daily routine adjustments to help manage fatigue and other symptoms.

Cognitive behavioural therapy is another effective method that aims to understand and improve the relationship between thoughts, feelings and behaviours.

Cognitive behavioural therapy is a therapeutic approach that aims to help individuals identify and modify harmful thought patterns that can worsen their emotional state.

For example, CBT techniques can help sufferers recognise thoughts that can exacerbate their anxiety and depressive symptoms. By working on these thoughts, patients can learn to develop more realistic and less harmful perspectives that improve their ability to cope with the illness.

Another important aspect of CBT is learning stress management strategies, such as mindfulness and relaxation techniques. These can help to reduce general stress levels, which can be exacerbated by the ongoing illness and associated uncertainty. Stress management is crucial as chronic stress can weaken the immune system and delay recovery.

Psychological support through counselling and CBT can not only help to overcome the emotional and cognitive challenges of Long COVID, but can also help to improve overall quality of life and well-being. By learning to deal more effectively with their thoughts and feelings and developing constructive coping strategies, sufferers can find a more active and self-determined way of dealing with their condition. Overall, integrating psychological support into the treatment plan for long COVID is essential to address the multidimensional challenges of these conditions.

f. Drug treatment

Drug therapies can play an essential role in the treatment of Long COVID, especially when it comes to alleviating the multiple and often distressing accompanying symptoms. As Long COVID can present with a range of symptoms, including fatigue, neurological impairment, sleep disturbances and chronic pain, treatment often requires a multidisciplinary approach, including the use of specific medications.

aa. Drug treatment of sleep disorders

Sleep disturbances are a common problem with Long COVID, which can significantly affect the overall recovery and quality of life of those affected. Treatment may include the use of sleep aids such as melatonin, which helps to regulate the natural sleep cycle, or prescription hypnotics such as zolpidem. However, it is important that these medications are used under careful medical supervision as they can promote dependence and may not be suitable for long-term use. Newer research approaches are also exploring the use of medications such as prazosin, which was originally developed to treat high blood pressure but has also shown positive effects on trauma-related sleep disorders, making it potentially useful for Long COVID patients with similar sleep problems.

bb. Drug treatment of pain

Chronic pain, especially muscle pain and headaches, are other common complaints of long COVID. Treatment may include non-steroidal anti-inflammatory drugs (NSAIDs) such as ibuprofen or naproxen, which are used to relieve inflammation and pain. In more severe cases, stronger painkillers, including opioids, may also be required, although these should be used with caution due to the risk of developing dependence. An alternative treatment option that is increasingly being researched is antidepressants such as duloxetine or amitriptyline, which are used to treat neuropathic pain and may also be effective in Long COVID by affecting central pain processing.

cc. Further research and approaches

More recent research into Long COVID is also focussing on the use of drugs to treat other aspects of the disease such as the persistent inflammation and autoimmune components. For example, immunomodulators and JAK inhibitors are being investigated that could regulate the overshooting immune system and dampen inflammatory processes. Such treatments could not only alleviate the symptoms, but also counteract the underlying mechanisms of the disease.

Research into the efficacy and safety of all these drugs is crucial, as the exact causes and mechanisms of Long COVID are not yet fully understood. Clinical trials and comprehensive research programmes are needed to develop effective and safe therapies for the diverse and often debilitating symptoms of Long COVID. In the meantime, personalised drug therapy tailored to the specific symptoms and needs of each patient remains an essential component of treatment strategies.

II Breathing difficulties

Breathing problems are one of the most challenging and worrying consequences that can occur in people recovering from COVID-19 or developing Long COVID. These respiratory symptoms range from mild shortness of breath on exertion to a constant feeling of breathlessness that can be present even at rest. In addition, many sufferers report a persistent feeling of tightness in the chest or difficulty taking a deep breath. These symptoms can have a profound impact on quality of life and daily functioning and pose a significant challenge for both patients and healthcare providers.

a. Description of the symptoms

The breathing problems experienced by those affected can take various forms:

- Shortness of breath (dyspnoea): This is the feeling of not being able to get enough air. It can occur during light physical activity, such as walking fast or climbing stairs, and is often so severe that even talking or eating can become strenuous.
- Chronic tightness in the chest: This feeling can be both frightening and physically uncomfortable and contributes to a constant feeling of discomfort.
- Difficulty taking a deep breath: This can include the feeling that each breath is incomplete, which can lead to frequent yawning or sighing in an attempt to take a deep breath.

b. Effects on daily life

The effects of these breathing problems on daily life are far-reaching.

- Restriction of physical activity: The ability to perform regular physical activities is severely restricted. Simple tasks that were previously effortless, such as walking short distances or light household chores, can become a challenge. This leads to a further reduction in physical fitness, as those affected are less active.
- Impairment of social and occupational participation: The restrictions can be so severe that those affected have to reduce or even give up their professional activities. Social interactions also become stressful, which can lead to social isolation.
- Deterioration of mental health: Chronic breathing problems can exacerbate anxiety and depression. The constant worry about your own breathing and the fear of your condition worsening can lead to ongoing stress and psychological strain.

c. Treatment approaches

The treatment of these breathing problems requires a multifactorial approach, which often includes the following elements.

aa. Pulmonary rehabilitation

Pulmonary rehabilitation is a comprehensive treatment programme for patients suffering from chronic breathing difficulties, as is often the case with Long COVID. This programme is designed to improve lung function, increase breathing efficiency and ultimately improve the quality of life of those affected. It combines different therapeutic approaches to address the specific needs of each individual, focusing on strengthening the patient both physically and psychologically and enabling them to better manage their symptoms.

As part of pulmonary rehabilitation, participants learn special breathing techniques aimed at strengthening the respiratory muscles and improving breathing efficiency. These techniques can help to reduce the feeling of breathlessness and maximise oxygen exchange, which is particularly beneficial during activities that require increased physical exertion. In addition, the programme includes physical training components that are tailored to the patient's individual capacity and tolerance. The aim is to gradually increase general endurance and physical performance without overtaxing those affected.

Another essential part of pulmonary rehabilitation is the educational component, which aims to give patients a better understanding of their condition and inform them about different strategies to manage their symptoms. This includes information on how the lungs work, how respiratory diseases affect the airways and how behavioural changes, such as avoiding pollutants or properly managing environmental stimuli, can contribute to symptom control.

In addition, pulmonary rehabilitation often also offers psychosocial support, as chronic respiratory problems are often accompanied by anxiety, depression and social isolation. Psychological support and group therapy sessions can help patients to cope better with the emotional and psychological stress of their illness. Participation in a group also offers the opportunity to share experiences and receive support from others experiencing similar challenges.

The combination of these therapeutic elements in a co-ordinated rehabilitation programme can significantly contribute to alleviating symptoms, improving functionality and helping patients to lead a more active and fulfilling life. The successes of pulmonary rehabilitation have shown that such an interdisciplinary approach is an effective way to manage the long-term effects of respiratory diseases such as long COVID.

bb. Drug therapy

Drug therapy plays a crucial role in the treatment of breathing difficulties associated with Long COVID and other chronic respiratory diseases. The use of medications such as bronchodilators and steroids aims to dilate the airways and reduce inflammation to make breathing easier and improve the quality of life of those affected.

Bronchodilators are medications designed to quickly widen narrowed airways. They work by relaxing the muscles around the airways, providing immediate relief from symptoms such as breathlessness. These medications are particularly helpful for people suffering from spastic airways and can be used for both acute and long-term treatments. By widening the airways, bronchodilators improve airflow, allowing patients to breathe more effectively and be more active without getting tired quickly.

Steroids, especially inhaled corticosteroids, are another important class of drugs used in the treatment of airway inflammation. They help to control chronic inflammation of the airways, which is a common cause of breathing difficulties and increased mucus production. By reducing inflammation, steroids can help to reduce the frequency and severity of breathlessness and potentially reduce the need for emergency treatment.

In addition to treating the symptoms, anti-inflammatory medications can prevent long-term damage to the airways that could otherwise lead to further complications and a deterioration in lung function. These medications help to suppress the underlying inflammatory processes that lead to airway obstruction, thus slowing the progression of the disease.

The combination of these medications can be particularly effective and is often tailored to the specific needs of the patient. The exact dosage and combination of medications will depend on the severity of symptoms and the overall health of the patient. While bronchodilators provide quick relief, steroids work more slowly to achieve long-term control. Treatment often needs to be adjusted regularly to achieve optimal results and minimise side effects.

Overall, drug therapy enables improved respiratory function and increased tolerance to physical activity, allowing patients to be more active and lead a fuller life. Careful monitoring by healthcare professionals is crucial to ensure that treatment remains effective and that the patient's health is not compromised by potential side effects of the medication.

cc. Oxygen therapy

Oxygen therapy is another treatment option for patients suffering from severe respiratory impairment, often seen in advanced stages of COVID-19 or other severe respiratory diseases. This form of therapy is used to increase the oxygen supply in the blood and ensure that vital organs such as the heart and brain are adequately supplied with oxygen, which is crucial for maintaining basic bodily functions.

In oxygen therapy, patients are supplied with additional oxygen via various systems, often through a nasal cannula or a breathing mask. This treatment can be used intermittently or continuously, depending on the patient's needs. For some patients, oxygen therapy is only necessary during certain activities or while sleeping, while others may require a constant supply to maintain their oxygen saturation at a safe level.

The decision to use oxygen therapy is typically based on a careful assessment of the patient's oxygen saturation, which is measured by a device called a pulse oximeter. If the oxygen level in the blood falls below a certain critical level, this can lead to hypoxia, a condition in which the tissues are not supplied with sufficient oxygen. Hypoxia can cause serious and permanent damage to the organs and must therefore be treated immediately.

The regular use of oxygen can help to alleviate symptoms such as fatigue, breathlessness and cognitive impairment, which are often associated with a lack of oxygen. In addition, oxygen therapy helps to improve the quality of sleep and increase physical performance as the body is better supplied with the essential element oxygen.

Despite the benefits of oxygen therapy, there are also some challenges and precautions to consider. Long-term use of oxygen can lead to dehydration of the nasal mucosa and there is a risk of oxygen toxicity if used incorrectly. Therefore, careful monitoring and regular assessment by healthcare professionals is necessary to adjust therapy accordingly and avoid potential complications.

In addition, psychological support can be very important in managing the anxiety and stress associated with breathing problems.

III Cardiovascular problems

Cardiovascular symptoms are a common and worrying manifestation of Long COVID that can significantly affect the daily lives of those affected. These symptoms range from palpitations and rapid or irregular heartbeat to chest pain. A particularly challenging condition associated with these symptoms is Postural Orthostatic Tachycardia Syndrome (POTS), which is characterised by a significant increase in heart rate when moving from lying to standing and can often lead to dizziness, lethargy and in some cases even fainting.

a. Description of cardiovascular symptoms

Palpitations and irregular heartbeats are often the result of overactivity of the sympathetic nervous system, which is activated in response to stress or inflammation in the body. These symptoms can occur intermittently or constantly, depending on the severity and the individual physiological response of the person affected. Chest pain experienced by Long COVID patients can be alarming and requires careful medical evaluation to differentiate cardiac causes from other possible causes such as musculoskeletal pain or acid reflux.

POTS occurs when the autonomic nerve, which is responsible for regulating involuntary bodily functions, malfunctions. This leads to inadequate adaptation of the body to changes in position, resulting in an excessive increase in heart rate and inadequate blood circulation, especially when standing up. Symptoms can be exacerbated by prolonged standing and are often so severe that they severely limit the ability of those affected to stand or move without experiencing dizziness or fainting.

b. Effects on the quality of life

The impact of these cardiovascular symptoms on quality of life can be profound. Physical performance is often severely limited, affecting the ability of those affected to carry out everyday activities or participate in work. This can lead to a reduction in independence and social isolation.

In addition, these symptoms can cause significant anxiety and stress, as sufferers are concerned about the potential long-term consequences of these conditions on their heart health. The fear of potential cardiac events or other serious complications can lead to increased psychological distress, which in turn can exacerbate cardiovascular symptoms.

c. Long-term health risks

Long-term cardiovascular problems, such as those experienced by some people after COVID-19 infection, can have profound and serious health consequences. Continued stress on the cardiovascular system through persistent symptoms such as palpitations, irregular heartbeat and chest pain can increase the risk of chronic and potentially life-threatening heart disease over time.

The heart and blood vessels are designed to adapt to different stresses, but prolonged or chronic stress can lead to permanent damage. With prolonged overwork, such as that caused by sustained tachycardia or high blood pressure, the heart must work harder to pump blood efficiently throughout the body. This overwork can lead to a thickening of the heart muscles, a condition known as hypertrophic cardiomyopathy. This thickening can affect the heart's efficiency over time, reducing the heart's ability to pump blood effectively, which can lead to heart failure.

In addition, persistent cardiovascular problems can affect the structure of the heart vessels. Chronic inflammatory processes triggered by long-term stresses such as infections can lead to changes in the blood vessels that promote arteriosclerosis, i.e. hardening and narrowing of the arteries. This increases the risk of heart attacks and strokes, as narrowed or hardened arteries can restrict blood flow to vital organs.

Cardiac arrhythmias triggered by persistent stress factors on the cardiovascular system can also have serious consequences. If the normal rhythm of the heart is disturbed, this can lead to inefficient blood circulation, which impairs the supply of oxygen and nutrients to the body. In severe cases, such arrhythmias can lead to sudden cardiac death, especially if they are not diagnosed and treated.

The cumulative effect of these problems can not only affect quality of life, but also shorten life expectancy. It is therefore crucial that people with persistent cardiovascular symptoms following COVID-19 infection receive close medical monitoring and treatment. Early detection and treatment of cardiovascular disease play a crucial role in minimising the risk of serious long-term complications. Regular cardiological examinations, adapted therapies and possibly lifestyle-adapted measures are essential to protect heart health and minimise the risk of future cardiovascular events.

Given the complexity and potentially severe nature of these symptoms, comprehensive medical care is crucial. This should include a thorough diagnosis to determine the exact cause and severity of symptoms, followed by a personalised treatment plan. Monitoring and regular follow-ups are also important to track the development of the condition and adjust treatment as needed. In many cases, a combination of medication, lifestyle adjustments and in some cases rehabilitation may be required to manage symptoms and improve quality of life.

The physical symptoms of Long COVID are not only challenging in isolation, but often interact with each other and together exacerbate the functional limitations. For example, fatigue can worsen breathing difficulties, and cardiovascular problems can exacerbate general fatigue and malaise. These interactions contribute to a vicious circle that makes it difficult for those affected to return to normality and recover effectively.

D. Neurological and cognitive symptoms

Long COVID can also include significant neurological and cognitive symptoms, which for many sufferers are among the most debilitating aspects of their condition. Commonly reported problems include brain fog, headaches and sensory disturbances.

I. Brain fog (cognitive impairment)

Brain fog in long COVID is a condition that occurs as a result of COVID-19 disease and is characterised by a variety of cognitive impairments.

a. Symptoms

People who suffer from this phenomenon report a range of symptoms that can significantly impair their daily functioning. These include memory impairment, concentration problems, slower information processing and difficulties with multitasking. These symptoms are often described as the feeling of a dense "fog" in the head, which hinders clear thinking and quick mental reactions.

The effects of brain fog are far-reaching and affect both the professional and private lives of those affected. At work, this cognitive impairment can mean that those affected struggle to fulfil their work requirements. They may have problems coping with complex tasks, communicating effectively or remaining productive under time pressure. These changes can lead to reduced work performance and even jeopardise the job.

Brain fog is also very noticeable in everyday life. Everyday decisions that may have been made effortlessly before the illness can become a major challenge. The ability to run the household or manage family commitments can also be impaired. This often leads to an increased dependence on others, which can greatly affect the feeling of independence and self-determination.

Many sufferers report a deterioration in their quality of life. Constantly dealing with cognitive impairment and its consequences can lead to frustration, stress and, in some cases, social isolation. The psychological burden caused by persistent cognitive fog can also increase the risk of other health problems, including depression and anxiety.

b. Treatment approaches

Coping with brain fog requires a multidimensional approach that incorporates medical, psychological and everyday practical aspects. This approach aims to improve the quality of life of those affected and enable them to function more effectively in both their professional and private lives.

- A thorough medical evaluation is critical to identify and treat underlying causes of brain fog. Doctors may prescribe medications that alleviate certain symptoms, such as drugs to improve concentration or reduce fatigue. In addition, they can refer sufferers to specialists such as neurologists or psychiatrists who specialise in cognitive impairment.
- Cognitive rehabilitation, offered by professionals such as occupational therapists or neuropsychologists, can be very helpful. These therapies include specialised exercises aimed at strengthening memory, attention and executive functions. Such programmes are often tailor-made and take into account the specific needs and goals of the individual.
- A healthy diet, regular physical activity and sufficient sleep are fundamental aspects that can help to improve cognitive function. Stress management techniques such as meditation, yoga or mindfulness-based stress reduction can also be supportive, as stress can often exacerbate the symptoms of brain fog.
- It is often necessary to adapt working conditions to meet the requirements of the brain fogger. This can include reducing working hours, reorganising tasks or providing special work equipment. Employers can provide support here by offering flexible working hours or home office options.

- As brain fog can also be emotionally stressful, professional psychological support is important. Psychologists or psychotherapists can provide strategies to deal with frustration, anxiety and depression, which often occur as secondary effects of chronic health conditions.
- The role of family and friends cannot be underestimated. They provide emotional support, understanding and practical help that is crucial for coping with daily life. Social support can also include relatives attending medical appointments to better understand how they can support.

Overall, managing brain fog requires a coordinated effort between sufferers, healthcare providers, employers and the social environment. Recognising and managing these conditions are essential steps to help sufferers improve their quality of life and maintain their independence as much as possible.

II Headaches

Headaches that occur in the context of Long COVID pose a particular challenge for those affected. These headaches are often chronic and can vary in intensity and form, either taking on migraine-like characteristics or being experienced as pressure and tension headaches. The variability and severity of the symptoms make treatment and daily management complex.

a. Characteristics of headaches

Migraine headaches are typically characterised by pulsating or throbbing pain that is often concentrated on one side of the head. This type of headache can be accompanied by nausea, vomiting and extreme sensitivity to light and noise. Pressure and tension headaches, on the other hand, feel as if a band is being pulled tightly around the head. They are less intense than migraines, but can last longer and spread over the whole head.

The persistent presence of these headaches can be highly disabling. They affect daily routines as the pain episodes can be so intense that normal activities, such as work, social interactions and even basic self-care, are interrupted or made impossible. The ability to concentrate and work effectively often suffers significantly, which can negatively impact the professional performance and career opportunities of those affected.

Headaches often occur together with other sensory hypersensitivities, such as increased sensitivity to light (photophobia) and noise (phonophobia). These additional symptoms can further reduce quality of life and make simple everyday activities such as going shopping or working in brightly lit environments a challenge. In some cases, these sensitivities can be so pronounced that sufferers withdraw and experience social isolation.

b. Management and treatment

Treating this type of headache often requires a combination of medication, lifestyle adjustments and therapeutic interventions. Medication can range from conventional painkillers to specific migraine medications, depending on the type and severity of the headache. Lifestyle adjustments, such as introducing regular rest, reducing stress and using relaxation techniques, can also be helpful.

Due to the chronic nature and associated emotional stress, psychological support is of great importance. Stress management techniques, cognitive behavioural therapy and possibly counselling can help sufferers to cope better with the effects of their headaches.

Holistic treatment and support for people with long-COVID headaches is crucial to improving their quality of life and helping them to function better both at work and at home. The key is to take a personalised approach that takes into account the individual symptoms and needs of those affected.

III Sensory disorders

The sensory disorders that can also occur as part of Long COVID are diverse and affect basic human abilities, which has a profound impact on those affected. These disorders include changes in the sense of smell and taste, visual disturbances and an altered sensitivity to touch.

These impairments are not only disruptive on a physiological level, but also affect the psychosocial and emotional health of those affected.

a. Changes in the sense of smell and taste

Some Long COVID patients experience a complete loss of the sense of taste and smell, known as anosmia and ageusia. These conditions can be particularly distressing as smell and taste are essential components of food intake and the pleasure of eating. The loss of these senses can have a profound impact on quality of life, affecting the ability to enjoy or even safely consume food.

In addition, other patients suffer from parosmia or dysgeusia, in which the perception of odours and tastes is distorted. Familiar smells and tastes can suddenly become unpleasant or completely unrecognisable, further complicating everyday life and negatively affecting the eating experience. These sensory changes are not only irritating, but can also affect eating behaviour and social interactions, as mealtimes are often the focus of social gatherings.

b. Visual disturbances

Visual problems that occur as part of Long COVID include a range of symptoms that affect vision and can have a far-reaching impact on daily life. Blurred vision, double vision and hypersensitivity to light are some of the most common visual complaints experienced by those affected. These symptoms can significantly affect the ability to perform everyday tasks.

Blurred vision can make it difficult to read text or clearly recognise details in the environment, which can be problematic not only in everyday activities such as reading signs or operating household appliances, but also at work. Double vision exacerbates these problems as it disturbs spatial perception and makes it difficult to recognise objects or find your way around a room. Such visual disturbances can be particularly dangerous when driving or during other activities that require precise visual coordination.

Hypersensitivity to light, also known as photophobia, causes normal daylight or artificial light to be perceived as unpleasant or painful. This can lead to sufferers having to avoid bright environments, which can limit participation in social activities and lead to a reduction in quality of life. The constant need to be in dimmed or dark rooms can also promote depressive moods or exacerbate existing depressive symptoms.

Occupational performance can suffer particularly from these visual problems. Tasks that require accurate visual perception, such as working on a computer, driving a vehicle or operating machinery, can become increasingly difficult. This can lead to reduced work productivity and, in the worst case, jeopardise the professional future of those affected.

The treatment of these visual symptoms often requires a combination of medical care, special visual aids such as glasses or contact lenses and, if necessary, adjustments at work or in the home environment to optimise lighting conditions. In some cases, therapeutic intervention can also help to improve visual resilience or develop adaptation strategies that enable those affected to lead an active and fulfilling life despite their limitations.

c. Changed sensitivity to touch

Altered perception of touch, also known as tactile sensitivity, can manifest itself in many ways and has various causes. It can range from a slight tingling sensation to an intense feeling of pain. This condition can be caused by a variety of factors, including neurological diseases, disorders of the nervous system, injuries or even psychological conditions such as anxiety disorders. If the nerves responsible for sensation in the skin are damaged or dysfunctional, they can transmit the signals they normally transmit incorrectly. As a result, everyday touches that are normally perceived as harmless or even pleasant are perceived as painful or unpleasant.

For example, people with neuropathic pain caused by damaged nerves can be extremely sensitive to touch, a condition known as allodynia. People with diabetes can also experience similar symptoms due to diabetic neuropathy. On the other hand, reduced sensitivity, as with numbness, can mean that touch is not perceived at all or is muted, which in turn can increase the risk of unnoticed injury.

Such changes in perception can also affect everyday interactions. Hugging, holding hands or even wearing clothes can become unbearable. This can promote social isolation and significantly affect quality of life. In addition, professions that require regular physical contact, such as carers or physiotherapists, can be particularly affected. It is important that people who experience such symptoms seek medical advice, as early diagnosis and treatment can help to alleviate symptoms and improve quality of life.

d. Effects on safety and everyday life

The inability to recognise spoiled food or smell dangerous gases such as smoke can jeopardise personal safety. The ability to react appropriately to visual and tactile stimuli can also be limited, which poses further risks in everyday life.

Eating is a social activity and an important part of culture and personal identity. Disorders in the sense of smell and taste can impair social life and lead to a loss of eating pleasure, which in turn can reduce general well-being. In addition, the limitations in communication and interaction with the environment can lead to social isolation and emotional difficulties such as depression and anxiety.

e. Management and treatment

There are various treatment approaches that can range from medication to physical therapies to behavioural therapies, depending on the cause and severity of the sensory disorder.

The treatment of altered perceptions of touch is usually tailored to the specific cause and severity of the symptoms. Medication often plays a central role, especially when it comes to relieving pain or improving nerve function. For example, in neuropathic pain, medications such as anticonvulsants, which were originally developed to treat epilepsy, or antidepressants can be used to stabilise nerve conduction and modulate pain signals.

Physical therapies are also an important part of treatment, especially when the cause is physical injury or disease affecting the musculoskeletal system. Physiotherapy can help to improve function, reduce pain and promote mobility. Techniques such as heat or cold therapy, electrical stimulation or ultrasound can also help to relieve pain.

Behavioural therapies and psychological support are also crucial if the altered perception of touch is related to psychological factors or if it causes significant anxiety and stress. Cognitive behavioural therapy, for example, can help patients cope better with their pain by addressing dysfunctional thought patterns and behaviours that can contribute to pain intensification. Relaxation techniques, mindfulness exercises and stress management strategies are also important elements of these therapeutic approaches.

In addition, alternative forms of treatment such as acupuncture or aromatherapy can also be used to alleviate symptoms and improve general well-being. However, each treatment requires individual customisation and should be carried out under the supervision of professionals to ensure the best efficacy and safety. It is also important that patients are actively involved in their treatment process and regularly talk to their practitioners about progress and any adjustments to therapy.

The neurological and cognitive symptoms of Long COVID can occur in isolation or in combination with other physical and mental health problems. These symptoms often require specialised medical and neuropsychological interventions. Treatment may include medication for symptom control, cognitive rehabilitation and psychotherapeutic support to help sufferers develop strategies to cope with these persistent and often distressing symptoms.

E. Psychological and emotional symptoms, long-term consequences

The psychological and emotional effects of Long COVID are just as significant as the physical symptoms and can have serious long-term consequences for those affected. The most common psychological symptoms include anxiety, depression and post-traumatic stress disorder (PTSD), all of which can have a significant impact on quality of life.

I. Psychological and emotional symptoms

a. Anxiety and depression

Long COVID brings with it a variety of long-lasting symptoms that can have a significant impact on the daily lives of those affected. Among these symptoms, psychological symptoms such as anxiety and depression are particularly prominent, as they significantly reduce quality of life and make recovery more difficult.

The persistent feelings of anxiety and depression in long COVID patients can be exacerbated by various factors. One of the main reasons is the ongoing uncertainty about the course of the disease. Many sufferers experience unpredictable fluctuations in symptoms with no clear prognosis, leading to constant worry about their own health and future. In addition, the physical limitations associated with Long COVID, such as persistent fatigue, breathing difficulties and pain, can increase the feeling of helplessness and loss of control over one's own life.

The isolation that often results from the long periods of illness also plays a major role in exacerbating these psychological symptoms. Due to the need to rest and recuperate, as well as the ongoing risk of infection in the early stages of the disease, many patients withdraw socially. This can lead to loneliness and increase the feeling of isolation, which in turn favours depressive moods.

The treatment of anxiety and depression in Long COVID requires an integrative approach. Psychotherapy, particularly cognitive behavioural therapy, can be effective in helping to identify and change negative thought patterns that contribute to anxiety and depression. Medication may also be appropriate, especially if symptoms are severe.

Social support is also crucial. Involvement in a community, whether through virtual support groups or face-to-face contact with friends and family, can help break the feeling of isolation and provide emotional support. Rehabilitation therapies can also help to improve physical health and well-being, which in turn has a positive impact on mental health.

Overall, managing anxiety and depression in long COVID requires a comprehensive view of the physical, psychological and social aspects of the disease. Through a combination of medical, psychological and social support, sufferers can achieve better management of their symptoms and improve their quality of life.

b. Sleep disorders

Sleep problems are a common and distressing symptom of long COVID and can have a variety of effects on the health and daily well-being of those affected. The causes of sleep disorders in this context are complex and often intertwined.

On the one hand, the anxiety caused by uncertainty about the course of the illness and the long-term health consequences can lead to sleep disorders. Worrying about one's own health and the future can lead to increased stress, making it difficult to rest at night. This is often accompanied by an overactive mind, which makes it difficult for those affected to switch off and fall asleep.

Secondly, the physical symptoms of Long COVID, such as persistent pain, breathing difficulties or night-time coughing fits, can directly disrupt sleep. These physical symptoms can make it difficult to fall asleep or cause patients to wake up frequently during the night, which significantly impairs the quality of sleep.

In addition, the disturbed rhythm during the acute phases of illness can have long-term effects on sleep patterns. Many people find it difficult to return to a normal sleep-wake cycle after an illness, especially if they have been bedridden for a long time or their daytime activities have been severely restricted.

Poor sleep can result in a cascade of negative health effects. Daytime sleepiness is one of the most direct consequences of poor sleep, which can affect the ability to perform daily tasks and stay active. Cognitive impairments, such as concentration problems, memory difficulties or slowed thought processes, are also common consequences. These in turn can have a negative impact on work performance and social interactions. In addition, prolonged poor sleep can weaken the immune system and worsen overall physical health, which is particularly problematic as the immune system plays a key role in recovery from Long COVID.

The treatment of sleep disorders in Long COVID often requires a multimodal approach. This may include adjusting the sleep environment, establishing a regular bedtime routine, relaxation techniques before bedtime and, if necessary, medication support. Professional counselling or specialised sleep therapy may be appropriate for more severe cases. In some cases, treatment of the underlying anxiety or depression that is affecting sleep may also be required.

c. Emotional exhaustion

The constant stress caused by the illness and concerns about your own health can lead to emotional exhaustion, which makes daily functioning even more difficult. This has already been explained in detail.

II Long-term consequences such as PTSD

Post-traumatic stress disorder (PTSD) can occur after a variety of traumatic experiences, including serious health events such as a COVID-19 infection or related hospitalisation. These experiences can leave profound psychological scars that manifest themselves in various symptoms.

a. Symptoms

The typical symptoms of PTSD include intense, unwanted flashbacks to the traumatic event, which can feel as real as if the person is reliving the event. Nightmares are also common and can disrupt sleep, leading to a general lack of sleep and associated daytime sleepiness and irritability. These memories are often so distressing that the affected person begins to avoid situations or activities that could evoke memories of the trauma. This avoidance behaviour can lead to the person increasingly withdrawing from social interactions and avoiding activities that were previously perceived as pleasant.

In addition, persistent anxiety is a central feature of PTSD. This can manifest itself as constant nervousness, increased vigilance (hypervigilance) and excessive startle reactions. People with PTSD can also feel permanently unsafe or threatened, even in situations that are objectively safe.

The treatment of PTSD after a serious illness such as COVID-19 often requires an integrated approach. Psychotherapy, particularly methods such as cognitive behavioural therapy or Eye Movement Desensitisation and Reprocessing (EMDR) therapy, has been shown to be effective in addressing the distressing memories and avoidance behaviours. In some cases, medication may also be indicated to manage the symptoms.

Support from family and friends is also very important. Social support can break through isolation and promote recovery. Education about PTSD and recognising its symptoms can also help to encourage sufferers to seek professional help and promote understanding in the social environment. It is important that both sufferers and their families recognise that PTSD is a treatable disorder and that the first step to recovery is to seek support.

b. Effects

Post-traumatic stress disorder (PTSD) can have a profound impact on almost all aspects of a person's daily life. The disorder not only affects mental health, but can also severely impact physical health, social relationships, job performance and overall quality of life.

The ability to participate in daily life can be affected by PTSD in various ways. For example, hypervigilance and constant anxiety can make everyday situations, such as travelling on public transport or visiting crowded places, seem overwhelming and unbearable. This can lead to individuals with PTSD choosing to avoid such situations, which in turn limits their ability to participate in normal social or professional activities.

Social relationships can also suffer greatly from the symptoms of PTSD. The tendency towards avoidance behaviour and withdrawal can lead to isolation as sufferers withdraw from friends and family to avoid potentially triggering interactions. This isolation can in turn exacerbate depressive symptoms and increase feelings of loneliness. In addition, the irritability and difficulties in dealing with strong emotions that often accompany PTSD can lead to tension and conflict in relationships.

In addition, PTSD can hinder recovery from other long COVID symptoms. Mental stress and emotional distress have been shown to have negative effects on the immune system and other physical functions, which can slow down recovery. Stress can increase inflammatory processes in the body and thus exacerbate symptoms such as fatigue, pain and cognitive impairment that often occur with Long COVID.

The effective treatment of PTSD is therefore crucial, not only to alleviate the psychological symptoms, but also to support overall recovery and the restoration of quality of life. Approaches such as psychotherapy, medication and support from social networks are central to addressing the various aspects of the disorder and helping those affected to lead a more active and fulfilling life again.

c. Treatment of psychological symptoms

aa. Psychotherapy

Cognitive behavioural therapy (CBT) is one of the most widely used and evidence-based forms of therapy for the treatment of mental disorders such as depression, anxiety and PTSD. The core of this therapy consists of identifying and modifying unhealthy and destructive thought patterns that negatively influence the behaviour and emotions of those affected.

CBT is based on the assumption that our thoughts have a decisive influence on our emotional world. In people with depression, anxiety disorders or PTSD, certain thought patterns can lead to negative emotional reactions. For example, a person with depression may tend to interpret experiences negatively, leading to further feelings of hopelessness and despondency. Similarly, a person with PTSD may experience triggers through certain thoughts and memories that cause intense anxiety or flashbacks.

In CBT, therapists work closely with patients to recognise such dysfunctional thought patterns. This is often done through techniques such as journaling or specific tasks that encourage patients to observe and analyse their thoughts during certain situations. Once these patterns are identified, strategies are worked on to challenge them and replace them with more realistic and less harmful thoughts.

An essential part of CBT is learning coping strategies that enable those affected to deal with stress better, overcome anxiety and prevent relapses. This often includes relaxation techniques, mindfulness exercises and practising new behaviours that can be used in anxiety-inducing or depressive situations.

The effectiveness of CBT has been proven by numerous studies showing that it can lead to improvements in emotional well-being. By changing the way a person thinks and interprets their world, CBT can help to alleviate symptoms and improve overall wellbeing. In addition, CBT offers the benefit that the techniques and strategies learnt help individuals to cope with future challenges on their own, which promotes long-term mental health.

bb. Drug treatment

Antidepressants and anxiolytics are two main categories of medications commonly used to treat mental disorders such as depression and anxiety. These medications play an important role in psychiatric treatment as they can help correct the chemical imbalances in the brain associated with these conditions.

Antidepressants mainly act on neurotransmitters in the brain, particularly serotonin and noradrenaline, which play an important role in regulating mood and emotions. There are different classes of antidepressants, including selective serotonin reuptake inhibitors (SSRIs), serotonin-noradrenaline reuptake inhibitors (SNRIs), tricyclic antidepressants (TCAs) and monoamine oxidase inhibitors (MAOIs). SSRIs and SNRIs are the most commonly prescribed antidepressants due to their efficacy and comparatively favourable side effect profile.

Anxiolytics, including benzodiazepines and buspirone, are used specifically to treat anxiety disorders. Benzodiazepines work quickly to relieve the symptoms of anxiety and panic by increasing the activity of GABA, a neurotransmitter that dampens the activity of neurones. However, they can lead to dependence and tolerance with long-term use, which is why they are usually only recommended for short-term use or acute anxiety. Buspirone is an alternative that works more slowly but does not have a high dependency potential and may therefore be more suitable for the long-term treatment of anxiety.

The use of antidepressants and anxiolytics must be carefully monitored as they can have side effects, including fatigue, weight changes, sexual dysfunction and in some cases even worsening of symptoms. For this reason, it is important that these medications are prescribed under the supervision of a qualified healthcare provider and that their effects are regularly evaluated.

In the overall treatment of mental disorders, medication is often an essential part of a broader treatment plan that may also include psychotherapy, lifestyle changes and support from social networks. The combination of these approaches can provide the best results by addressing not only the chemical imbalances in the brain, but also the underlying psychological and social factors that contribute to the development and maintenance of the disorder.

cc. Support groups and social support

Talking to others who are going through similar experiences can be enormously helpful. Support groups provide a space to share experiences and learn strategies for coping with the illness. Regular physical activity, a healthy diet and sufficient sleep are also important to promote mental well-being and strengthen resilience to mental health problems.

Treating the psychological and emotional symptoms of Long COVID always requires a comprehensive approach that addresses both the mental and physical health needs of those affected. Through such holistic care, people with Long COVID can be better supported to improve their quality of life and recover from the extensive effects of the disease.

F. Diagnostic procedures

I. Diagnostic criteria of Long COVID according to WHO and CDC

The diagnostic criteria for Long COVID, also known as the post-COVID-19 condition, varies slightly between different health organisations such as the World Health Organisation (WHO) and the Centers for Disease Control and Prevention (CDC) in the USA. However, both organisations have developed guidelines to help doctors and healthcare professionals effectively identify and treat Long COVID. Here is an overview of the current WHO and CDC criteria:

a. World Health Organisation (WHO)

The WHO defines Long COVID (post-COVID-19 state) as the onset of symptoms that usually begin 3 months after SARS-CoV-2 infection and last for at least 2 months without an alternative diagnosis that can fully explain the symptoms. The most important aspects of the WHO definition are

- Symptoms should start at least 3 months after the initial COVID-19 illness and last at least 2 months.
- Symptoms may be similar to those of the acute COVID-19 phase or may be new symptoms that

appear after recovery. These include fatigue, shortness of breath, cognitive dysfunction and others that affect daily functioning.

- There should be no alternative diagnosis that fully explains the symptoms.

b. Centres for Disease Control and Prevention (CDC)

The CDC defines post-COVID conditions as a wide range of new, recurring or persistent health problems that people may experience more than four weeks after the initial viral infection. The CDC highlights:

- Symptoms include persistent fatigue, headaches, head fog, sleep disturbances, persistent fever, muscle aches, and more.
- The CDC recognises that symptoms and their severity can vary from person to person and that anyone who has had a COVID-19 infection can develop persistent or late symptoms, regardless of the severity of the initial symptoms.
- The CDC emphasises the need for an individualised approach to the management of long COVID, including a comprehensive assessment and tailoring of treatment plans to patients' specific symptoms and needs.

Both organisations, WHO and CDC, recognise that Long COVID is challenging to diagnose as there are no specific tests that can clearly identify the condition. The diagnosis is therefore mainly based on the clinical assessment of symptoms and the patient's medical history, including the exclusion of other causes that could explain the symptoms. This underlines the complexity of the post-COVID-19 condition and the need for careful and individualised medical care.

II Tests that are used to identify long COVID symptoms

The diagnosis of Long COVID is mainly based on the clinical assessment of a patient's persistent symptoms following COVID-19 infection, as there are no specific tests developed exclusively to identify Long COVID. However, various tests and investigations are used to assess symptoms, rule out other causes and determine the extent of involvement of different organ systems.

a. Blood tests

Monitoring inflammatory markers and performing organ function tests are essential aspects in the diagnosis and monitoring of many diseases, including those associated with chronic inflammation or systemic disorders.

Inflammation markers such as C-reactive protein (CRP) and erythrocyte sedimentation rate (ESR) are particularly valuable for recognising inflammatory processes in the body. CRP is a protein that is produced by the liver and whose level in the blood can rise during acute inflammation. An elevated CRP level can indicate a variety of inflammatory conditions, from infections to autoimmune diseases. The erythrocyte sedimentation rate measures how quickly red blood cells in a blood sample sink to the bottom within an hour. A higher rate can be an indicator of inflammation, as inflammatory processes can increase the rate of cell sedimentation.

Other biomarkers that can be helpful in the assessment of inflammation include pro-inflammatory cytokines such as interleukin-6 (IL-6) and tumour necrosis factor-alpha (TNF-alpha), which are directly involved in the inflammatory response. The measurement of these markers can be particularly useful in the diagnosis and monitoring of autoimmune diseases or chronic inflammatory processes.

Organ function tests play a critical role in assessing overall health and identifying potential damage or dysfunction in major organs. Liver function tests, such as alanine aminotransferase (ALT) and aspartate aminotransferase (AST), can provide information about liver damage or disease. Kidney function tests, including creatinine and blood urea nitrogen (BUN), help to assess how well the kidneys are filtering waste products from the blood.

Electrolyte tests, which measure sodium, potassium, chloride and other important ions, are important to monitor the balance of these essential nutrients in the body, which is critical to maintaining normal cell function and fluid balance. A complete blood count (CBC) provides information about the health of blood cells and can reveal anaemia, infections and other blood disorders.

Regular monitoring of these tests can be crucial in detecting disease development early, assessing the effectiveness of treatments and ruling out other potential health issues. This is particularly important for those with chronic conditions or those experiencing persistent symptoms such as Long COVID, as it allows health conditions to be closely monitored and adjusted accordingly.

b. Imaging procedures

Imaging techniques such as chest X-rays, CT scans and MRI are crucial to gain deeper insights into possible organic damage and changes in the body following an illness such as COVID-19. These techniques are particularly valuable in the diagnosis and management of complications that can result from severe infections.

Chest X-ray and CT scan of the thorax are helpful imaging techniques for evaluating the lungs and other structures in the chest. After a COVID-19 infection, these images can be used to identify persistent or new lung problems. They can reveal signs of fibrotic changes, persistent inflammation or other complications such as pulmonary embolisms, which can occur in COVID-19 patients after severe courses. Compared to an X-ray, a CT scan offers a much more detailed view of the lung structure and can detect even the smallest changes that may not be visible on an X-ray.

MRI of the brain is another important examination method, especially when it comes to neurological complications. In patients who experience persistent neurological symptoms such as headaches, cognitive impairment, dizziness or sensory disturbances after COVID-19 infection, an MRI can be used to rule out structural or pathological changes in the brain. This high-resolution imaging technique makes it possible to create detailed images of brain tissue to identify inflammation, haemorrhage, ischaemia or other neurological abnormalities. By using MRI, doctors can better understand the extent to which the nervous system is affected and initiate appropriate therapeutic measures.

These imaging techniques are therefore not only critical for diagnosis, but also for the management and monitoring of disease progression and recovery after severe infectious diseases. They provide valuable information that is crucial for planning and implementing targeted treatment tailored to the patient's specific needs and conditions.

c. Cardiological examinations

The ECG (electrocardiogram) and the echocardiogram are two basic diagnostic tools in cardiology that contribute significantly to the assessment of heart health.

An ECG is a non-invasive procedure that measures the electrical activity of the heart. It is particularly useful for recognising cardiac arrhythmias. By recording the electrical impulses that occur during each heartbeat, doctors can identify different types of arrhythmias as well as other heart problems such as heart attack or ischaemia. An ECG can be performed quickly and provides immediate results, allowing doctors to respond quickly to potentially life-threatening conditions.

An echocardiogram is an ultrasound examination of the heart that provides detailed images of the heart's structure and function. The echocardiogram can assess the movement of the heart muscle, the function of the heart valves and the general pumping function of the heart. This is particularly important in patients who have symptoms such as chest pain, palpitations, shortness of breath or fatigue. By evaluating aspects such as ejection fraction - a measure of the amount of blood the heart ejects with each beat - the echocardiogram helps diagnose conditions such as heart failure, valve disease and other cardiac abnormalities.

Both tests are crucial for cardiovascular diagnostics and play an important role in the early detection of heart problems, especially in patients who have experienced a serious illness such as COVID-19. COVID-19 has been linked to a range of cardiovascular problems, including myocarditis, arrhythmias and even acute myocardial infarction. Early and accurate assessment of cardiac function by ECG and echocardiogram can therefore be crucial to prevent long-term damage and initiate appropriate treatment.

d. Pulmonary function tests

Spirometry is an essential diagnostic procedure in pulmonology that is used to assess the function of the lungs. This test measures how much and how quickly a person can inhale and exhale air, providing valuable insight into the presence and extent of lung dysfunction.

During spirometry, the patient is asked to take a deep breath and then exhale as forcefully and quickly as possible into a mouthpiece connected to a spirometer. This device records both the volume of exhaled air and the speed of exhalation. The most important measured values that are determined during spirometry are

- FEV1 (forced expiratory volume in one second): This is the amount of air a person can expel within the first second of forced expiration. A reduced FEV1 value can indicate an obstruction of the airways, as occurs in diseases such as asthma or chronic obstructive pulmonary disease (COPD).
- FVC (forced vital capacity): This is the maximum total amount of air that can be exhaled after a deep inspiration. A reduction in FVC can indicate a restrictive lung disease in which the lungs cannot be completely filled with air, as can be the

case with pulmonary fibrosis or after a severe COVID-19 infection.

- The ratio of FEV1 to FVC is also a critical indicator: a low ratio typically indicates obstructive lung disease, while normal or high values with reduced FVC may indicate restrictive disease.

Spirometry is particularly useful to diagnose, classify and monitor the presence and severity of respiratory diseases. It can also help to monitor the effectiveness of treatments and guide adjustments to therapy. For patients who have been through an infection such as COVID-19 and report persistent symptoms such as breathlessness, spirometry can be crucial in determining whether there is long-term damage to the lungs and how best to treat it.

e. Neurological and cognitive assessments

aa. Cognitive tests

Neuropsychological tests are specialised assessment procedures used to assess the extent and nature of cognitive impairment. These tests are particularly valuable in diagnosing and monitoring conditions that affect cognitive functioning, such as the condition known as "brain fog", which is often reported by patients with Long COVID or other neurological conditions.

Brain fog, as described in detail above, is a colloquial term for symptoms that include reduced cognitive function, such as concentration problems, memory difficulties, confusion and reduced mental clarity. These symptoms can have a significant impact on everyday life as they make simple tasks and decisions difficult.

The appropriate neuropsychological testing includes a variety of tests that measure different aspects of cognitive function, including:

- Memory: Tests such as learning and recognising lists or stories evaluate short and long-term memory.
- Attention and concentration: Tasks that require rapid responses to visual or auditory stimuli assess the ability to focus and maintain attention.
- Executive functions: Tests that measure planning, problem solving and cognitive flexibility help to assess how well a person can manage and perform complex tasks.
- Language: Vocabulary and fluency tests can reveal whether there are any impairments in verbal communication.
- Visual-spatial skills: Tasks that assess the understanding and manipulation of visual information are important in determining how well someone can deal with spatial relationships.

These tests are typically conducted by a neuropsychologist and can take several hours to complete. The results provide a detailed representation of a person's cognitive profile and help to identify specific cognitive deficits. They are crucial for the development of individualised treatment plans that can target specific cognitive weaknesses, and they provide a basis for monitoring disease progression and response to treatment.

For patients who experience cognitive impairment after a disease such as COVID-19, neuropsychological testing can provide crucial insights into the impact of the disease on the brain and help to take appropriate therapeutic or rehabilitative measures to improve cognitive function and thus quality of life.

bb. Neurological examination

A complete neurological examination is a basic tool used by neurologists to assess the presence and extent of neurological deficits. This examination is crucial for the diagnosis and management of neurological disorders, as it provides a comprehensive assessment of various functions of the nervous system.

During a neurological examination, the doctor assesses several key areas that together provide a complete picture of a person's neurological health:

- Mental status test: This includes the assessment of cognitive functions such as alertness, orientation, attention, memory, language skills and logical thinking. These tests provide information about how well the brain processes and stores information.

- Cranial nerve testing: There are 12 cranial nerves that control different areas such as vision, facial muscles, hearing, taste and swallowing. Testing these nerves can provide information about potential problems such as impaired vision, facial paralysis or loss of taste.
- Motor function: The assessment includes muscle strength, tone and coordination. The doctor may instruct you to perform simple tasks such as pushing against resistance or performing specific movements to identify weakness or coordination problems.
- Reflex tests: These involve testing involuntary reflexes, which provide important information about the integrity of the nervous system. Abnormal reflexes can indicate problems in the central or peripheral nervous system.
- Sensitivity test: This assesses whether the person feels pain, touch, temperature and vibration normally. Sensitivity disorders can indicate damage to the nerve pathways that transmit sensory information from the body to the brain.
- Gait and balance tests: The doctor can observe how the person walks, stands and keeps their balance. These tests are important to determine whether there are musculoskeletal disorders or problems with the sense of balance.

This examination provides important information that, together with other diagnostic tests such as imaging (e.g. MRI of the brain) and laboratory tests, helps to create a comprehensive picture of a person's neurological condition. Based on the results of the neurological examination, the doctor can recommend specific treatments or order further investigations to precisely determine the causes of the deficits identified and address them accordingly.

cc. Further specialised tests

Autonomic function tests are specialised tests designed to assess the function of the autonomic nervous system. The autonomic nervous system controls unconscious bodily functions, including heart rate, digestion, breathing patterns and blood pressure regulation. Dysfunctions in this system, known as dysautonomia, can cause a variety of symptoms and conditions, including postural orthostatic tachycardia syndrome (POTS), in which there is a significant increase in heart rate when standing up.

Some of the commonly performed tests to assess autonomic nerve function include:

- Tilt table test: This is one of the main methods used to diagnose POTS and other forms of orthostatic intolerance. During this test, the patient is placed on a special table, which is then tilted from a horizontal to a vertical position to observe how the body, particularly the cardiovascular system, reacts to the change in gravity. Doctors measure the heart rate and blood pressure to determine whether there are any unusual rises or falls.

- Quantitative Sudomotor Axon Reflex Test (QSART): This test measures the ability of the nerves to control the sweat glands, which is an indicator of autonomic nerve function. In this test, small electrodes are applied to the skin to stimulate and measure sweat production, which allows conclusions to be drawn about the function of the sympathetic nervous system.

- Valsalva manoeuvre: This test requires patients to inhale deeply and then attempt to exhale while the mouth and nose are closed, creating pressure in the chest. This manoeuvre tests the response of the cardiovascular system to changes in pressure, reflecting autonomic control of heart rate and blood pressure.

- Heart rate variability analysis (HRV): This test evaluates the heart's ability to change rate in response to respiratory cycles. Reduced variability may indicate dysautonomia.

These tests are also important for the diagnosis and management of dysautonomia and related disorders. They help to identify the specific functions and dysfunctions of the autonomic nervous system, which in turn enables targeted treatment aimed at alleviating symptoms and improving the quality of life of those affected.

These tests not only help in the diagnosis of Long COVID, but also in the development of an individualised treatment plan tailored to each patient's specific needs and symptoms. As Long COVID can affect a variety of organ systems, a multidisciplinary approach is often required to effectively manage the different aspects of the condition.

III. Challenges and limitations in diagnosis

The diagnosis of Long COVID poses a significant challenge due to various factors. These challenges and limitations affect both clinical practice and research. Here are some of the key issues that arise in the diagnosis of Long COVID.

a. Variability of symptoms

Long COVID is a complex disease characterised by a variety of symptoms that persist or reappear after recovery from an acute COVID-19 infection. The broad symptomatology and the variability of symptoms pose significant challenges for both patients and healthcare professionals.

Long COVID can affect multiple organ systems, resulting in a wide range of symptoms. Among the most common are persistent fatigue, often severe enough to interfere with daily functionality, and respiratory problems, which can range from shortness of breath to persistent coughing. Cognitive impairment, often described as "brain fog", includes problems with memory, concentration and the ability to perform complex tasks. In addition, neurological disorders such as headaches, dizziness and sensory changes may occur. Other symptoms can include cardiovascular problems, joint and muscle pain and changes in mood and emotional well-being.

The symptoms of Long COVID are not only varied, but also often fluctuating, meaning that they can vary in intensity and occur periodically. Some patients experience phases in which the symptoms subside and flare up again, often without any recognisable triggers. This inconsistency makes it very difficult to develop a standardised diagnostic criterion and to evaluate the effectiveness of treatments.

The treatment of Long COVID must therefore be highly individualised and may include a combination of symptomatic therapies, rehabilitative measures and psychosocial support. The aim is to improve the quality of life of those affected and help them cope with the ongoing effects of their condition. Given the complexity of Long COVID, ongoing research is needed to better understand the mechanisms underlying this condition and to develop more effective treatment strategies.

b. Lack of specific tests

The diagnosis of Long COVID is a significant challenge due to the lack of specific biomarkers and the complex nature of the disease. The clinical picture is characterised by a variety of symptoms that can affect numerous organ systems and is complicated by the lack of clearly defined diagnostic criteria.

There are currently no specific biomarkers that reliably indicate Long COVID. This means that there are no simple tests, such as blood tests, that could confirm a Long COVID diagnosis by their results alone. Instead, the diagnosis relies predominantly on the patient's detailed symptom history and the long-term course of these symptoms following COVID-19 infection. Doctors must also rule out other potential causes for the symptoms, which often requires extensive investigations to rule out conditions such as autoimmune diseases, chronic fatigue syndromes or neurological disorders.

Even when abnormalities are detected in diagnostic tests, their attribution to Long COVID is often problematic. Many of the symptoms and test results observed in Long COVID, such as inflammatory markers, changes in lung function tests or abnormalities in imaging, can also occur in a variety of other medical conditions. This makes it difficult to clearly attribute these test results to Long COVID and requires careful interpretation by experienced physicians.

The difficulty in diagnosing Long COVID and the potential overlap with other conditions requires a multidisciplinary approach to the treatment and care of those affected. Doctors from different specialities, such as pulmonologists, cardiologists, neurologists and psychiatrists, often work together to gain a comprehensive understanding of the patient's condition and develop a personalised treatment plan.

In addition, further research is urgently needed to better understand the pathophysiological mechanisms underlying Long COVID, identify specific biomarkers and develop more effective diagnostic criteria and treatment methods. These efforts will not only help to improve medical care for affected patients, but will also broaden the overall understanding of these long-term consequences of COVID-19.

c. Overlap with other diseases

The diagnosis of Long COVID proves to be particularly complex, not only because of the variety and fluctuation of symptoms, but also because of the need to rule out other medical conditions and consider the role of co-morbidities.

Long COVID is often diagnosed as a diagnosis of exclusion. This means that doctors must first rule out other illnesses that could cause similar symptoms. These include chronic fatigue syndrome (CFS), fibromyalgia, various autoimmune diseases, neurological disorders and mental illnesses such as depression or anxiety disorders. Each of these conditions can cause symptoms such as fatigue, pain, cognitive impairment and sleep disturbances, which are also common in Long COVID. The process of elimination often requires an extensive number of tests and examinations, including blood tests, imaging, neurological examinations and sometimes psychiatric evaluations. This can not only be time-consuming, but also a source of frustration for patients seeking a quick and clear diagnosis.

The situation is further complicated if patients have pre-existing medical conditions. Many people suffering from Long COVID may have had health problems such as diabetes, heart disease, chronic lung disease or immune disorders prior to their COVID-19 infection. These pre-existing conditions can not only exacerbate the symptoms of Long COVID, but also complicate its management. For example, medications used to treat Long COVID symptoms such as pain or inflammation could interact with medications already used to treat the pre-existing conditions. In addition, existing health problems can delay recovery and increase the risk of further complications.

The care of long COVID patients therefore requires an individualised approach that takes into account both the unique symptoms and existing comorbidities. A multidisciplinary approach is often necessary to address both the physical and psychological aspects of the condition. This usually involves a combination of medical treatment, physical therapy, psychological support and sometimes social support to help patients improve their quality of life and cope with the long-term effects of COVID-19.

d. Subjective nature of some symptoms

The challenges of diagnosing and treating Long COVID are further complicated by the fact that many of the symptoms, such as fatigue and cognitive impairment, are highly subjective. These subjective symptoms cannot be directly measured by standard tests such as blood tests or imaging techniques, which complicates the diagnostic process.

For symptoms such as fatigue, "brain fog" (cognitive impairment), pain and other sensory disturbances, the diagnosis is mainly based on the patient's description. As these symptoms cannot be measured objectively, a careful and comprehensive medical history is crucial. The treating physician must collect detailed information about the nature of the symptoms, their onset, duration, intensity and impact on the patient's daily life. This can be supported by specific questionnaires and symptom assessment scales that help to quantify the severity and specific characteristics of the symptoms.

The role of the doctor is crucial here. Doctors must not only be empathetic and attentive to their patients' concerns, but also be able to read between the lines to understand the often elusive and changing symptoms of

Long COVID. The challenge is to put these subjective descriptions into the context of other, measurable medical information and rule out other possible causes for the symptoms.

In addition, doctors often need to differentiate between physical and psychological aspects of the symptoms, as Long COVID can have both physical and emotional effects. This often requires interdisciplinary collaboration with specialists such as neurologists, psychiatrists, psychologists and physiotherapists to get a comprehensive picture of the patient's health and develop an effective treatment strategy.

The treatment of Long COVID therefore requires not only an individualised approach tailored to the specific needs and symptoms of each patient, but also ongoing assessment and adjustment of treatment plans. This underlines the need for patient-centred medical practice that takes patients' subjective experiences seriously and places them at the centre of the diagnosis and treatment processes.

e. Lack of recognition and understanding

The challenges in dealing with Long COVID are compounded not only by the nature of the disease itself, but also by systemic and organisational barriers. Inconsistent definitions and the varying availability of resources are two of the main obstacles that need to be overcome to ensure effective care for Long COVID patients.

A major problem in the treatment of Long COVID is the lack of a standardised definition and diagnostic criteria. Different health organisations and countries have developed their own guidelines, which may differ in the description of symptoms, the duration of symptoms after infection and the inclusion criteria for diagnosis. These differences lead to inconsistencies in diagnosis, which in turn complicates research and understanding of the disease. Patients may receive different diagnoses and treatment recommendations depending on where they seek medical help. This can also affect the comparability of study data, slowing down the development of effective treatment strategies.

Another problem is the lack of specialised resources in many medical facilities. Despite the growing awareness of Long COVID, many areas lack specialised clinics dedicated solely to the treatment of this condition. In addition, medical staff are often not adequately trained in recognising and treating the complex and varied symptoms of Long COVID. As a result, many patients do not receive the support they need to deal with the long-term effects of their condition. The lack of specialised resources can also result in patients having to endure long waiting times for treatment or their symptoms not being fully recognised and addressed.

Coordinated efforts at national and international level are needed to overcome these challenges. Clear, standardised guidelines for the diagnosis and treatment of Long COVID, based on the latest scientific evidence and regularly updated, are needed. In addition, better training of medical staff on Long COVID is crucial, as is the creation of specialised treatment centres that are accessible to affected patients. These measures would not only improve the quality of patient care, but also help to deepen the understanding of this still relatively new disease.

These challenges highlight the need for further research to better understand the pathophysiology of Long COVID, identify specific biomarkers and develop effective management strategies. At the same time, broader recognition and education in the medical community is needed to ensure appropriate care for the growing number of people with Long COVID.

G. Treatment approaches to combat Long COVID

The comprehensive treatment and care of patients with Long COVID requires a multidisciplinary approach involving different healthcare providers to effectively manage the multiple and often complex symptoms. Here is a more detailed look at the roles that different healthcare professionals play in the care of Long COVID patients:

I. General practitioners

They are often the first point of contact for patients and play a crucial role in the initial assessment and monitoring of symptoms. GPs not only coordinate primary care, but also manage the flow of information between patients and specialists. If necessary, they refer patients to appropriate specialists to ensure that all aspects of Long COVID are carefully addressed.

II Specialists

- Infectious disease specialists and internists help manage complex cases by providing specialised knowledge about infectious diseases and their long-term effects.

- Pulmonologists are crucial in the diagnosis and treatment of persistent breathing problems, a common complaint in Long COVID.
- Cardiologists take over the monitoring and treatment of cardiovascular problems that occur in some Long COVID patients.
- Neurologists address neurological and cognitive impairments that can range from mild disorders to severe dysfunction.
- Psychiatrists and psychologists are important for the treatment of mental health, which can be affected by the long-term effects of COVID-19.

III Therapists

- Physiotherapists develop customised exercise programmes that help to rebuild strength and endurance, especially after long periods of immobility.
- Occupational therapists provide support in regaining the skills for everyday activities and offer practical solutions for the management of limitations in everyday life.
- Speech therapists are important for patients who have communication difficulties or cognitive impairments.

IV. Nutritionist

They play an important role by offering nutritional plans specifically designed to boost the immune system, reduce inflammation and promote overall health.

V. Carers and social support

- Carers provide daily support and long-term care, which can be particularly crucial for patients with more severe cases of Long COVID.
- Social workers help navigate the healthcare system, assist with bureaucratic hurdles and provide access to community resources and support networks.

This team-based approach ensures that Long COVID patients receive comprehensive care that is tailored to their individual needs to improve their quality of life and support their recovery.

The coordination of these different disciplines is crucial for the effective treatment of Long COVID. A central aspect of the multidisciplinary approach is regular communication between the professionals involved to adjust treatment plans and respond to changes in the patient's condition. This integrative approach aims not only to alleviate physical symptoms, but also to improve patients' psychological resilience and quality of life.

VI Drug therapies, their effectiveness and possible side effects

The drug treatment of Long COVID is complex and depends heavily on the specific symptoms and their severity. As Long COVID has a wide range of symptoms that can affect multiple organ systems, drug therapy is often symptom-orientated. Here are some of the common drug therapies, their effectiveness and possible side effects.

a. Medication to relieve fatigue and boost energy levels

Stimulants such as methylphenidate: These can help relieve fatigue and improve cognitive function. Side effects can include nervousness, sleep disorders and palpitations.

b. Medication for neurological and cognitive symptoms

Antidepressants (SSRIs, SNRIs): They can be helpful for depression, anxiety and some neuropathic pain. Side effects can include weight changes, sexual dysfunction and sometimes increased anxiety or agitation at the start of treatment.

Acetylcholinesterase inhibitors: In cases of cognitive impairment, these drugs can help to improve memory

function. Side effects can include gastrointestinal complaints and headaches.

c. Medication for the treatment of respiratory problems

Inhaled corticosteroids and bronchodilators: These may
be prescribed for persistent respiratory symptoms to improve lung function. Side effects may include oral thrush
(fungal infection in the mouth) and hoarseness.

d. Cardiovascular medication

Beta-blockers: These can be used in patients with POTS
(postural orthostatic tachycardia syndrome) or other
cardiac arrhythmias to control the heart rate. Side effects
can include tiredness, cold hands and feet and sleep disturbances.

ACE inhibitors and ARBs: In patients who develop hy-
pertension or other cardiovascular complications after
COVID-19, these medications may be useful for blood
pressure control. Side effects may include dizziness, hy-
perkalaemia and impaired renal function.

e. Antiviral and immunomodulatory therapies

Interferons and other immunomodulators: In some
cases, these drugs can be used to modulate the immune

response, especially if persistent viral activity or an autoimmune component is suspected. Side effects may include flu-like symptoms and worsening fatigue.

The effectiveness of these and other therapies can vary greatly, and treatment often needs to be personalised based on the patient's individual symptoms and responses. It is important that the treatment of Long COVID takes place under the supervision of physicians who have experience with the diverse and often changing aspects of these conditions. Long-term studies and further research are needed to determine the optimal treatment strategy for Long COVID patients and to understand the efficacy and safety of therapies.

VII. Non-drug therapies such as physiotherapy, occupational therapy and special breathing exercises

The comprehensive treatment of Long COVID through non-drug therapies represents a holistic approach that aims to improve the quality of life of those affected and help them cope with the diverse and often distressing symptoms. These approaches include physical, cognitive, emotional and social aspects of the disease and offer patients a wide range of supportive measures. Here is an overview of the most important non-medication therapies:

a. Physiotherapy

Physiotherapy is a cornerstone in the treatment of Long COVID, especially because of its effectiveness in restoring mobility and strength. Individually adapted exercise programmes, which are gradually increased in intensity and scope, help to improve physical endurance and functional ability. For example, a progressive walking programme that starts with short walks and is gradually intensified can be crucial in restoring cardiovascular and muscular health.

b. Occupational therapy

Occupational therapy helps patients to manage daily activities effectively. By training in energy conservation techniques and adapting the home environment, patients are helped to increase their independence and avoid fatigue. Occupational therapists develop strategies specifically tailored to patients' needs and limitations to improve their quality of life.

c. Special breathing exercises

Breathing exercises are essential for improving lung function and breathing capacity. Techniques such as diaphragmatic breathing promote breathing efficiency and can help to reduce the symptoms of breathlessness. The use of devices such as PEP (Positive Expiratory

Pressure) can also support the airways and make breathing easier.

d. Cognitive rehabilitation

Cognitive rehabilitation addresses memory, attention and problem-solving deficits through targeted exercises and strategies. This can help to improve the cognitive function and daily functioning of patients with brain fog and other cognitive impairments.

e. Psychological support

Psychological and psychiatric support is crucial to overcome the emotional and psychological challenges of Long COVID. Therapies such as cognitive behavioural therapy can effectively address negative thought patterns and help develop strategies for stress management and mindfulness.

f. Nutritional therapy

A customised diet rich in vitamins, minerals and antioxidants can strengthen the immune system and reduce inflammatory processes in the body. Nutritionists can create specific diet plans that are tailored to the individual health needs of patients.

g. Social support

Social support through self-help groups and social services provides a platform for exchange and mutual help. This helps to reduce feelings of isolation and provide practical help with everyday challenges.

By integrating these non-drug therapies into the Long COVID treatment plan, comprehensive and personalised care can be ensured, taking into account all aspects of the disease and maximising the patient's recovery and quality of life.

VIII Psychological support services and behavioural therapies for coping with anxiety and depression

Psychological support services and behavioural therapies play a crucial role in treating the psychological effects of Long COVID, particularly anxiety and depression. These forms of therapy aim not only to alleviate symptoms, but also to provide sufferers with strategies to better manage their conditions and improve their quality of life. Here are some important psychological support services and behavioural therapies.

a. Cognitive behavioural therapy

CBT is one of the most effective methods for treating anxiety and depression. It is based on the realisation that dysfunctional thought patterns and beliefs can lead to psychological problems. CBT helps patients to recognise, question and ultimately change their negative thought patterns.

CBT includes techniques such as cognitive restructuring, where patients learn to replace negative thoughts with more realistic ones, and behavioural experiments that help to address anxiety in controlled steps.

CBT has been shown to be particularly effective in reducing depression and anxiety disorders and is well documented in the scientific literature.

b. Mindfulness-based therapies

These therapies integrate mindfulness exercises to help people experience the moment without judgement. This can be particularly helpful in breaking the cycle of anxiety and depressive thoughts.

Techniques include meditation, breathing exercises and body scan exercises aimed at developing a deeper awareness of one's own body and mind.

Studies show that mindfulness-based approaches can reduce symptoms of anxiety and depression and contribute to general emotional regulation.

c. Interpersonal psychotherapy

IPT is a short-term form of therapy that focuses on interpersonal relationships and social functioning to treat mental health problems.

IPT helps patients to improve their social skills, resolve conflicts in relationships and seek support from others.

IPT has been shown to be particularly useful in the treatment of depression by helping people to improve their social interactions and relationships.

d. Group therapy

Group therapy provides a platform for people experiencing similar mental health problems to share their experiences and find support.

Under the guidance of a therapist, participants can share their experiences, learn from others and develop strategies together to overcome their challenges.

The group dynamic can reduce the sense of isolation that often accompanies depression and anxiety, and promote a sense of community and understanding.

e. Exposure therapy

This method is often used for anxiety disorders, especially when specific fears or phobic reactions are present.

It involves the gradual and controlled confrontation with the source of anxiety in order to reduce the anxiety reaction.

Exposure therapy can be very effective in reducing anxiety symptoms by helping patients overcome their fears in a safe setting.

These psychological therapies can be used in isolation or in combination, depending on the patient's specific needs and circumstances. For people with Long COVID, it is particularly important that the chosen form of therapy is flexible and adapted to changing health conditions. Ongoing therapeutic support is often necessary to help those affected to cope with the effects of their condition in the long term.

IX. Importance of nutrition, adequate fluid intake and adapted physical training

A healthy diet, adequate fluid intake and adapted physical exercise are fundamental elements for overall health and play a particularly important role in the recovery and management of long COVID symptoms. These aspects not only contribute to physical recovery, but also improve mental health and overall well-being. Here is a detailed look at the importance of each of these elements:

a. Importance of nutrition

A nutrient-rich diet supports the immune system. Vitamins and minerals such as vitamin C, vitamin D, zinc and selenium play a central role in the functioning of the immune system.

Foods with anti-inflammatory properties, such as omega-3 fatty acids in oily fish, nuts and seeds, and antioxidants in colourful fruits and vegetables, can help reduce systemic inflammation, which can play a role in Long COVID.

A balanced diet helps to regulate energy levels, which is particularly important as many Long COVID patients suffer from persistent tiredness and fatigue.

b. Importance of fluid intake

Adequate fluid intake is crucial for overall body function. Dehydration can exacerbate fatigue, headaches and cognitive impairment, symptoms that are common with Long COVID.

Water helps to remove waste products and toxins from the body, which supports kidney function and can help to improve well-being.

c. Importance of adapted physical training

Adapted training can help to overcome the muscle weakness and atrophy caused by the disease. It also improves cardiovascular health, which is particularly important as COVID-19 can affect the cardiovascular system.

Regular physical activity is known to lead to the release of endorphins, the so-called "happiness hormones", which help to alleviate depression and anxiety.

As ability and endurance vary from person to person, especially with long COVID, the training programme should be individually adapted to avoid overexertion and promote recovery.

H. Current research and developments on Long COVID

Long COVID research is a rapidly growing and active field as scientists around the world work to understand the causes, mechanisms and effective treatments for this complex disease. Here is an overview of some key ongoing research projects and clinical trials aimed at better understanding and treating Long COVID:

I. RECOVER Initiative (USA)

The RECOVER project, initiated by the National Institutes of Health (NIH), represents a major scientific effort to study the long-term effects of SARS-CoV-2 infection. This initiative aims to develop an in-depth understanding of Long COVID by focusing on extensive research into the physiological, psychological and biological aspects of the disease.

A key focus is on identifying biological markers and risk factors that are crucial for the development of Long COVID. Researchers in the RECOVER programme are striving to decipher the mechanisms behind the persistent and often debilitating symptoms experienced by many sufferers. They are also investigating how this disease develops and changes over time in order to develop more effective treatment strategies and intervention methods tailored to specific patient groups. Through this broad research initiative, it is hoped to make significant advances in the treatment and management of Long COVID, which should ultimately significantly improve the quality of life of those affected.

II PHOSP-COVID study (United Kingdom)

The PHOSP-COVID study, led by the University of Leicester, is a crucial research initiative in the UK focussing on the long-term health impact of COVID-19, specifically on patients discharged from hospital. This extensive study aims to gain in-depth insights into the after-effects of COVID-19 disease to gain a better understanding of the long-lasting consequences of the disease.

A central focus of the PHOSP-COVID study is to understand which factors influence the recovery of patients. This includes analysing demographic, biological and lifestyle-related factors that could potentially have a positive or negative impact on the recovery process. The researchers are also trying to determine how the coronavirus affects the organs of those affected in the long term. Particular attention is being paid to possible permanent damage to the lungs, heart, kidneys and other organs, as well as long-term neurological and psychological consequences.

In addition, the study aims to develop recommendations on how to improve medical care for patients suffering from the long-term effects of COVID-19. This includes the optimisation of treatment strategies and rehabilitation measures that are specifically tailored to the needs of long-COVID patients.

The results of this study are of great importance, as they should not only deepen the medical understanding of Long COVID, but also provide concrete approaches to sustainably improve the quality of life of those affected and effectively support their recovery.

III COVID-LTI (Long-Term Impact of Infection) study (Europe)

This European study provides a comprehensive approach to investigate the long-term impact of COVID-19 on patients treated both in hospital and at home. This study is particularly important as it covers a wide range of patient experiences and attempts to paint a detailed picture of the after-effects of the disease.

A key aim of the study is to understand why some patients develop long-lasting symptoms, often referred to as Long COVID. Researchers are focussing on identifying the biological and perhaps genetic factors that contribute to these persistent health problems. This includes investigations into inflammatory processes, immune responses and possibly persistent viral reservoirs in the body.

The study also looks at how COVID-19 affects different organ systems over time. This includes observing long-term effects on the cardiovascular system, the lungs, the neurological system and other important areas of the body. The aim is to recognise patterns or common complications resulting from the initial viral infection and to document these systematically.

Another important aspect of the study is the socio-economic impact of Long COVID. This involves analysing how the persistent symptoms affect patients' ability to work, their quality of life and their social interactions. This includes assessing the financial burden on patients and the healthcare system as a whole, as well as analysing how different socioeconomic groups are affected differently by Long COVID.

Through this multi-faceted approach, the study hopes to address not only the medical but also the social challenges of Long COVID and ultimately develop strategies that enable effective support and treatment for all those affected.

IV. DECIPHER (UK)

The DECIPHER study represents a significant scientific approach to exploring the role of genetic factors in the development of Long COVID. By understanding the genetic basis, scientists can better understand why some people experience long-lasting symptoms after a COVID-19 infection, while others recover completely.

A key aim of the DECIPHER study is to identify genetic patterns associated with increased susceptibility to Long

COVID. Researchers are analysing the DNA of patients who show long-term consequences after COVID-19 infection to detect common genetic variants or abnormalities that distinguish this group from those who experience no or only minor after-effects.

By identifying specific genetic markers, the study hopes to be able to make precise predictions about which individuals are at higher risk of developing Long COVID. This would make it possible to identify risk groups at an early stage and initiate preventive measures to minimise the severity of long-term symptoms.

In addition to risk prediction, the study aims to use the findings on genetic factors to develop personalised treatment approaches. Understanding genetic predispositions could enable doctors to design more specific and effective therapies for long COVID patients based on their individual genetic profile.

The DECIPHER study could thus contribute significantly to the development of personalised medicine in the treatment of Long COVID by not only showing who is most at risk, but also which treatments are most promising for different genetic profiles. This knowledge could enable targeted, efficient and patient-specific treatment aimed at significantly improving the quality of life of those affected and accelerating recovery.

V. Long COVID-19 studies in Australia

In Australia, several studies are dedicated to the long-term effects of COVID-19, known as Long COVID. These research projects focus in particular on the areas of cognitive rehabilitation and return to work. The central objective of these studies is to develop and evaluate effective rehabilitation strategies. These strategies should help those affected to regain their ability to work and achieve an overall improvement in their quality of life.

Given that COVID-19 can leave many people with long-lasting symptoms, such as fatigue, difficulty concentrating and other cognitive impairments, the development of targeted rehabilitation programmes is of great importance. These programmes are designed to provide individually tailored therapies and support mechanisms that enable those affected to gradually return to their professional and social lives.

The Australian research teams often work on an interdisciplinary basis to develop a comprehensive understanding of long COVID and test therapies for their effectiveness. These include, for example, physiotherapy interventions, cognitive behavioural therapy and specialised counselling services aimed at addressing the specific needs of Long COVID patients. In addition, the effectiveness of work rehabilitation programmes is also being researched, which are designed to gradually reintegrate people into the work process in a way that is adapted to their individual capabilities.

These and many other studies play a crucial role not only for the patients themselves, but also for the country's healthcare system and economy, as successful rehabilitation of long-COVID patients helps to reduce long-term healthcare costs and improve labour productivity. The results of this research could also serve as a model for similar challenges internationally in the future.

VI Future therapeutic approaches that are currently under development

Given the growing importance and complex nature of Long COVID, researchers worldwide are working to develop innovative and effective therapeutic approaches. These future therapies include a variety of approaches, including potential drugs, vaccines and other treatment strategies that target the underlying mechanisms and symptoms of Long COVID. Here are some promising areas of research.

a. Antiviral medication

Research into antiviral medication in the context of Long COVID is particularly relevant as some of the persistent symptoms may be caused by continued viral activity in the body. There is evidence that in some Long COVID patients, the coronavirus remains active in certain cells of the body, causing long-term health problems. This assumption has led scientists to evaluate the efficacy of existing antiviral drugs in this specific context.

The current stage of development of these studies is focussed on finding out to what extent these drugs can effectively combat the SARS-CoV-2 virus or reduce its activity in the body. These drugs, which were originally developed to treat other viral infections such as influenza or HIV, have the potential to inhibit viral replication and thus lead to a reduction in the viral load and associated symptoms of long COVID.

The potential of these antiviral drugs is significant. If the ongoing studies confirm that these drugs can effectively reduce viral activity in long COVID patients, this could lead to a significant improvement in the quality of life and health of this patient group. Successful use of antiviral drugs could also mean that the duration and severity of the disease is effectively reduced, which in turn would allow a faster return to normal activity and a reduction in the long-term consequences of COVID-19.

However, this research is still in its infancy and extensive clinical trials are needed to ensure the safety and efficacy of antiviral treatment specifically for Long COVID. Given the complex nature of Long COVID and the different symptoms experienced by patients, it is also important to develop personalised treatment plans that are tailored to the specific needs and medical backgrounds of those affected.

The research and development of antiviral drugs for the treatment of COVID-19 and specifically Long COVID is being driven by a large number of pharmaceutical companies worldwide. Some of the best-known companies active in this field include

- Gilead Sciences: The manufacturer of Remdesivir, one of the first drugs approved specifically for the treatment of COVID-19. Remdesivir has antiviral properties aimed at preventing the replication of the coronavirus.
- Pfizer: In addition to its mRNA vaccine, Pfizer is also developing antiviral therapies against COVID-19. They are working on an oral drug called Paxlovid, which is intended to inhibit the activity of the virus in combination with another active ingredient.
- Merck & Co. (known as MSD outside the USA and Canada): Merck has developed molnupiravir, an oral antiviral drug created in collaboration with Ridgeback Biotherapeutics. Molnupiravir was developed to stop the spread of coronavirus in infected individuals.
- Roche: In collaboration with Atea Pharmaceuticals, Roche is working on antiviral drugs specifically targeting SARS-CoV-2.

These companies are currently at the forefront of research and development of drugs that have the potential to revolutionise the treatment of COVID-19, including use in patients with long COVID.

b. Immunomodulators

Research on immunomodulators in the treatment of Long COVID is an emerging field that focuses on regulating the immune system to prevent or treat excessive inflammatory responses that may contribute to Long COVID. Many Long COVID symptoms, including fatigue, joint pain and neurological complaints, are associated with persistent inflammatory processes in the body.

Immunomodulators are substances that are able to influence the immune system. They can either enhance or suppress the immune response, depending on how they are formulated. In the case of long COVID, the use of immunomodulators could aim to modulate the immune response so that it is not too strong and thus damages the body.

The companies and research groups working on the development of immunomodulators for Long COVID include some well-known names from the pharmaceutical and biotech industries. These research approaches include:

- Cytokine inhibitors: Some research is focussing on drugs that inhibit specific cytokines that play a role in inflammatory processes. These drugs, which are often used in the treatment of autoimmune diseases such as rheumatoid arthritis, could potentially be adapted for long COVID patients.

- Janus kinase (JAK) inhibitors: This class of drugs is also being investigated for its potential to modulate inflammatory processes in long COVID. JAK inhibitors could help to interrupt the signalling pathway that leads to inflammation.
- Interleukin blockers: These drugs aim to block specific interleukins that are involved in the inflammatory response. Research in this direction could show how modulation of these molecules could improve the symptoms of Long COVID.

The potential of these immunomodulators lies in their ability to dampen the inflammatory response and thus improve a variety of symptoms associated with long COVID. By reducing inflammation, the overall disease burden for patients could be reduced, leading to an improved quality of life. However, it should be noted that research in this area is still in its infancy and requires extensive clinical trials to confirm the safety and efficacy of these therapies.

This development of immunomodulators, particularly for use in long COVID, involves numerous biotechnology and pharmaceutical companies. Some of these companies are already known for their work on immunomodulators in other medical fields and are now expanding their research into the potential treatment of COVID-19 and its long-term consequences. Notable companies and their research initiatives include:

- Roche: This pharmaceutical company has experience in the development of immune modulators and is active in research for COVID-19-related treatments. Among other things, Roche has developed tocilizumab, which was originally used for rheumatoid arthritis and is now also being tested in the treatment of severe COVID-19 cases.

- Regeneron Pharmaceuticals: Known for its monoclonal antibodies used to treat COVID-19, Regeneron is also researching other immunomodulatory therapies that have the potential to modulate the inflammatory responses caused by the virus.

- Novartis: Novartis has a wide range of immunomodulators used in various therapeutic areas and is exploring their potential applications in relation to COVID-19 and Long COVID. Their experience with interleukin inhibitors could be particularly relevant here.

- Pfizer: In addition to its successful COVID-19 vaccine, Pfizer is also researching treatments that can modulate the immune response to alleviate the symptoms of Long COVID.

These companies are part of a larger network of research institutions and smaller biotech companies that are also active in this field. Research into immunomodulators is complex and often requires partnerships between different players in the healthcare sector to develop innovative solutions.

c. Neuropathic painkillers

Research into drugs to treat neuropathic pain in the context of Long COVID is particularly important as many sufferers complain of persistent pain syndromes that significantly affect their quality of life. Neuropathic pain caused by nerve damage or dysfunction can occur in Long COVID patients due to nerve impairment caused by the virus.

Research in this area is still in its infancy, as Long COVID itself is a relatively new condition and the scientific community is still in the process of fully understanding the pathophysiology. Drugs currently used to treat neuropathic pain, such as antiepileptic drugs (e.g. gabapentin, pregabalin), certain antidepressants (e.g. amitriptyline, duloxetine) and local anaesthetics (e.g. lidocaine patches), are being investigated in clinical trials to determine whether they are also effective and safe for Long COVID patients.

However, the possibility of adapting existing drugs to treat neuropathic pain for Long COVID offers great potential. These drugs could help to alleviate specific symptoms such as chronic pain, burning, tingling and muscle spasms. By reducing these symptoms, sufferers could experience a significant improvement in their daily functioning and overall quality of life. In addition, the successful treatment of neuropathic pain in long COVID patients could also help to reduce the use of opioids, which are often associated with a high risk of side effects and dependence.

Research in this area requires close collaboration between neurologists, pain specialists, virologists and other medical professionals to ensure that treatment approaches are not only effective but also safe for this particular patient group. As Long COVID encompasses a variety of symptoms that can affect multiple body systems, a multidisciplinary approach is crucial to provide the best care and most effective therapies.

Various pharmaceutical companies that already have experience in the development of drugs for neuropathic pain are also researching potential applications of their products for long COVID patients.

- Pfizer: Pfizer manufactures drugs such as pregabalin (Lyrica), which is used to treat neuropathic pain. Given their broad portfolio and strong presence in COVID-19 research, they could also investigate the efficacy of their existing painkillers for Long COVID.
- Lilly: Eli Lilly produces duloxetine (Cymbalta), an antidepressant that is also prescribed for the treatment of diabetic neuropathy and other forms of neuropathic pain. They may be evaluating the use of this medication for long COVID pain management.

- Novartis: As one of the largest pharmaceutical players in the world, Novartis has access to resources and technologies to research the efficacy of its pain management products in new therapeutic areas such as Long COVID.
- Teva Pharmaceuticals: Teva, known for its generic and specialised drug products, including those for pain relief, could potentially be involved in exploring the application of existing pain therapies to long COVID symptoms.

These companies and others in the pharmaceutical industry are potentially able to repurpose their existing drugs or develop new drugs specifically for the treatment of long COVID-related neuropathic pain.

d. Vaccines against Long COVID

Research into the role of COVID-19 vaccines in reducing Long COVID symptoms or even preventing this long-term condition is a relatively new and rapidly developing field. While the primary role of COVID-19 vaccines is to prevent the disease, scientists are increasingly investigating how these vaccines could also affect the risk of Long COVID.

Researchers around the world are exploring the potential mechanisms by which vaccines could have a positive impact on Long COVID. One theory is "immune reset", where vaccination modulates the immune system to correct persistent or misdirected immune responses that cause Long COVID symptoms. This research is ongoing and data from clinical trials and observational studies are continuously being collected to test these hypotheses.

Early data from various studies suggest that people who have been fully vaccinated against COVID-19 are less likely to develop Long COVID if they contract the virus. In addition, there are anecdotal reports and preliminary study results suggesting that some people already suffering from Long COVID have experienced an improvement in their symptoms after vaccination. These observations have raised hopes that vaccination could serve not only as a preventive measure but also as part of the treatment strategy for Long COVID.

Leading pharmaceutical companies and research institutions that have developed COVID-19 vaccines are involved in these studies.

- Pfizer/BioNTech and Moderna: Both companies have developed mRNA vaccines and are actively researching the effects of these vaccines on Long COVID.
- AstraZeneca and Johnson & Johnson: These companies, which produce vector-based vaccines, are also investigating the potential benefits of their products in the context of Long COVID.

Investigating how vaccinations affect Long COVID is important for understanding the disease and could inform future treatment approaches. Such findings could help to develop targeted therapies for Long COVID based on modulation of the immune system.

e. Cell-based therapies

Research into stem cell-based and other cell-based therapies as potential treatments for Long COVID is also an exciting and innovative field based on the principles of regenerative medicine. These therapies utilise the ability of stem cells and other specialised cells to repair or regenerate damaged tissue and potentially modulate the immune system.

Research into stem cell-based therapies for long COVID is largely in the early stages, with some studies already in preclinical or initial clinical phases. These approaches are promising as stem cells have the ability to differentiate into different cell types and thus potentially repair a variety of tissue damage caused by the virus. In addition, it is being investigated whether stem cells have anti-inflammatory properties that could help to reduce the chronic inflammatory processes associated with long COVID.

The potential therapeutic benefits of these cell-based approaches include

- Tissue repair. The ability to regenerate damaged tissues such as lung or heart muscle tissue could be crucial, as many Long COVID symptoms result from such damage.

- Immune modulation: By modulating the immune response, the persistent inflammation that leads to Long COVID symptoms could be reduced and the immune system brought into balance.
- Improvement in quality of life: The reduction in symptoms such as fatigue, breathing difficulties and neurological impairments could significantly improve quality of life.

companies and research institutions that are leaders in this field:

- Athersys: A biotechnology company specialising in the development of stem cell-based therapies and potentially expanding their application to Long COVID.
- Mesoblast: A company focussing on the development of mesenchymal stromal cells (MSCs), which are known for their anti-inflammatory and regenerative capabilities.

Research into cell-based therapies for long COVID is still in its infancy and further studies are needed to determine their safety, efficacy and best practices. Due to the complexity of these therapies, regulatory hurdles are high and clinical development can be lengthy. Nevertheless, these advanced treatments offer the hope of novel approaches to treating the often difficult-to-treat and diverse symptoms of Long COVID.

These and other research approaches are ultimately crucial to understanding the underlying causes of Long COVID and developing effective treatments. Given the complexity of the syndrome and the different symptoms it encompasses, a wide range of therapies are likely to be required to meet the different needs of those affected.

VII Gaps in current research and the need for further studies

Long COVID research is still in its infancy, and although significant progress has been made, there are still many gaps and unanswered questions that require further investigation. These gaps in current research emphasise the need for additional studies to gain a more complete understanding of Long COVID and develop more effective treatment strategies

a. Understanding the pathophysiology

The limited understanding of the pathophysiology of Long COVID poses a challenge in medical research and treatment. Despite numerous ongoing studies and research initiatives, the question of why some people develop long-term symptoms while others make a full recovery remains largely unanswered. These gaps in knowledge mean that the exact mechanisms by which the virus causes long-lasting health problems are not yet fully understood.

Knowledge gaps in the pathophysiology of Long COVID are as follows

- Immune response: A persistent misdirected or excessive immune response could contribute to the persistence of long COVID symptoms. Understanding how the immune system responds after an acute infection and potentially leads to chronic conditions is key.
- Autoimmunity: There are hypotheses that Long COVID may have autoimmune-like traits, in which the immune system mistakenly attacks the body's own cells after the virus is no longer detectable.
- Viral persistence: In some patients, the virus may remain latent in certain body cells and trigger chronic inflammatory processes.
- Vascular damage: COVID-19 is also associated with vascular damage, which can lead to circulatory disorders and resulting tissue damage.

In order to understand the mechanisms behind Long COVID, comprehensive biological and medical studies are needed to investigate various aspects of the disease:

Molecular biological and immunological research is needed to understand the interactions of the virus with the human immune system and how these can lead to persistent symptoms.

Long-term observational studies are needed to follow the trajectories of recovery or persistent symptoms in different patient groups.

Comparative studies between people who fully recover and those who develop long COVID could be informative in identifying risk factors and protective mechanisms.

This research is crucial to develop effective treatment strategies and ultimately identify preventive measures that can minimise the risk of developing Long COVID. The multidisciplinary approach to research, involving virology, immunology, genetics and other medical specialities, is essential. The knowledge gained could not only be relevant for the treatment of Long COVID, but also provide deeper insights into other post-viral and chronic inflammatory diseases.

b. Identification of biomarkers

The identification of specific biomarkers for Long COVID is a critical area of research that is crucial for the diagnosis and management of this disease. The current challenge is that Long COVID presents with a wide range of symptoms that can vary widely between patients and are often non-specific. This makes early identification and effective treatment of affected individuals difficult.

The importance of biomarker research for Long COVID is due to the following reasons:

- Early diagnosis: Specific biomarkers could help doctors to diagnose Long COVID quickly and accurately, even in patients with mild initial COVID-19 symptoms or those who have never tested positive for the virus.
- Assessment of severity: Biomarkers could also provide information on the severity of the disease, which would be valuable for adapting treatment plans and for prognostic assessments.
- Monitoring treatment success: Furthermore, they could serve as tools to monitor the success of therapies and, if necessary, make adjustments if patients do not respond to treatments as expected.

Comprehensive clinical studies are needed to identify and validate potential biomarkers. These studies should include a broad range of patients to cover the diversity of Long COVID symptoms.

Research should be interdisciplinary, with immunologists, virologists, endocrinologists and other specialists working together to understand the complex interactions between the virus and different body systems.

The collection of samples in biobanks and the use of advanced data analysis methods are crucial for recognising patterns that could indicate specific biomarkers.

Examples of potential biomarker research areas are

- Inflammatory markers: Studies could focus on identifying specific inflammatory markers associated with Long COVID.
- Autoantibodies: Since there is a possibility of an autoimmune component in Long COVID, the identification of autoantibodies could be helpful.
- Metabolic and proteomic signatures: Research could also investigate metabolic changes or specific proteins in the blood that are associated with Long COVID.

The successful identification and validation of biomarkers for long COVID would not only improve diagnosis and treatment, but also deepen our understanding of this complex disease and potentially contribute to the development of targeted therapies.

c. Development and evaluation of treatments

The challenge faced by healthcare professionals in the treatment of Long COVID is the absence of specific, evidence-based treatment guidelines. Long COVID presents with a variety of different and often interconnected symptoms that can vary from patient to patient. This makes standardised treatment difficult and requires an individually tailored approach.

The development of evidence-based guidelines is crucial to ensure that patients with Long COVID receive the most effective and safest treatment. Without such guidelines, clinicians often have to rely on their clinical experience or on treatment approaches developed for other conditions with similar symptoms. This can lead to inconsistent treatment outcomes and a delay in recovery.

Urgent need for clinical studies:

- Diversity of therapeutic approaches: Clinical trials are necessary to evaluate the efficacy and safety of different treatment approaches. These include drug therapies that aim to treat specific symptoms such as inflammation, pain or neurological complaints. Non-drug therapies such as physiotherapeutic interventions that focus on restoring physical function and reducing fatigue, as well as psychological support to cope with the mental and emotional stresses associated with Long COVID, also need to be investigated.
- Interdisciplinary approaches: Research should also consider the development of interdisciplinary treatment models that address the complex nature of Long COVID. This could include teams of physicians from different specialities, physiotherapists, psychologists and other health specialists.
- Patient-specific treatments: Studies should aim to develop personalised treatment plans tailored to patients' individual symptoms and needs.

Based on the results of clinical trials, comprehensive treatment guidelines should be developed that can serve as a guide for doctors worldwide. The medical community must be informed about new treatment strategies and research results through continuous education programmes.

The development and validation of treatment approaches through rigorous clinical research is critical to improving the quality of life of long COVID patients and effectively supporting the healthcare system. Such studies are complex and require coordination and cooperation between research institutions, hospitals and other medical facilities to create a broad data base that provides a solid foundation for future treatment recommendations.

d. Diversity and inclusion in studies

The lack of demographic diversity in clinical trials, particularly in the context of Long COVID, is a significant issue that can limit the generalisability and relevance of research findings. Studies that have a homogeneous group of participants run the risk of overlooking important differences in symptoms, disease progression and treatment responses that are influenced by demographic factors such as age, gender, ethnicity and socio-economic conditions.

Involving a broad range of participants in clinical trials is crucial to ensure that research findings are comprehensive and applicable to different populations. This is particularly important in Long COVID as the disease affects a broad and diverse population. Different demographic groups may react differently to the virus, experience different side effects and suffer different long-term consequences. Strategies to improve demographic diversity in studies are as follows:

- Recruitment strategies: Researchers should develop targeted recruitment strategies to ensure that study participants are drawn from diverse demographic groups. This can be done through partnerships with community organisations, clinics in diverse geographic regions and the use of multilingual study materials.
- Methods of analysis: Studies should be designed to collect and analyse data that allows demographic differences in outcomes to be identified. This includes collecting comprehensive data on ethnicity, gender, age, socioeconomic factors and other relevant variables.
- Consideration of cultural and social factors: Researchers need to consider cultural, social and economic factors that may influence participation in studies and health outcomes. This includes the design of study protocols that are accessible to participants from different backgrounds.
- Interdisciplinary teams: Forming research teams that include experts from different disciplines

and cultural backgrounds can help bring a broader perspective to research and encourage the development of studies that take demographic diversity into account.

By implementing these strategies, researchers can ensure that the results of their studies reflect the realities of a diverse global population. This is critical to developing effective, safe and equitable treatment strategies for long COVID that address the needs of all populations.

e. Long-term consequences and their management

The long-term management of Long COVID and its impact on healthcare systems are critical areas that have not been adequately researched. This gap in research leads to uncertainties in the treatment of patients and complicates the planning and adaptation of healthcare services to meet the demands of this new and complex disease. Longitudinal studies play a key role as they allow patients to be followed over longer periods of time, providing valuable insights into the development and progression of the disease.

Conducting longitudinal studies helps researchers understand the dynamics of Long COVID, including the factors that may influence improvement or worsening of symptoms. This type of research is fundamental to capturing how the disease affects patients' health over time and which treatment strategies are most effective. Such insights are important not only for direct patient care, but also for strategic planning within healthcare systems. They enable better resource planning and allocation, the adjustment of insurance benefits and the development of policies that address the long-term needs and challenges of long COVID patients.

Furthermore, the inclusion of a broad and diverse patient population in these studies enhances the accuracy and relevance of the research findings. By including different demographic groups, scientists can gain more comprehensive and universal data that will enable a better understanding of the differential impact of Long COVID on different population segments. International collaborations reinforce this approach by enabling an exchange of knowledge and experience across different health systems, improving the global response to the Long COVID problem.

The implementation of such research initiatives is crucial not only to improve the medical care and quality of life of those directly affected, but also to minimise the burden that Long COVID places on healthcare systems worldwide. This will ultimately help ensure that healthcare systems are better prepared for these and future health challenges.

f. Psychological and social effects

The psychosocial impact of Long COVID is another significant research gap that urgently needs to be addressed. Many sufferers experience significant challenges that affect not only their physical health, but also their social and psychological well-being. The long-term effects such as inability to work, social isolation and mental health problems can have a profound impact on patients' lives and therefore require comprehensive consideration and targeted research efforts.

The need for further research in this area is crucial to capture a complete picture of the psychological and social impact of Long COVID. This includes the development and evaluation of support systems specifically designed to meet the diverse and complex needs of those affected. The focus is not only on direct medical treatment, but also on the provision of psychological counselling, social support and rehabilitation services that help patients to resume their social roles and improve their quality of life.

To effectively address the psychological effects of Long COVID, it is important to pursue interdisciplinary research approaches that involve experts from psychology, social work, medicine and other relevant disciplines. Such teams can develop comprehensive treatment and support plans that are tailored to the specific psychosocial impact of the disease. In addition, large-scale epidemiological studies can provide valuable data to help understand the prevalence and nature of psycho-social problems associated with Long COVID.

Another important aspect of research should be the collection of long-term data to track the duration and progression of psychosocial effects. This is particularly relevant as some mental health problems such as depression or anxiety disorders can only manifest themselves over time and become chronic.

The development of effective support systems and therapeutic interventions specifically tailored to the needs of long COVID patients will not only improve individual coping, but also help to reduce the overall burden on families and society. Ultimately, such comprehensive research and the resulting interventions can help make the healthcare system more resilient and responsive to similar challenges in the future.

Closing these research gaps is crucial to unravelling the many unknown aspects of Long COVID and ultimately developing effective treatments and support for the millions of people affected worldwide.

Challenges and outlook

Long COVID not only presents significant medical and health challenges, but also has an impact on social and economic levels. The broad and multi-layered effects require a comprehensive approach in order to understand the full consequences of the disease and develop suitable countermeasures.

The social and economic challenges posed by Long COVID are complex and require coordinated action by various actors such as governments, health organisations and civil society. Long COVID leads to an ongoing burden on healthcare systems, as those affected often require long-term medical care, which can lead to overburdened healthcare facilities and divert resources from other essential medical services. In addition, many people with Long COVID experience a significant reduction in their social interactions due to their symptoms. The physical limitations and persistent fatigue make it difficult to participate in social life, which can lead to isolation and loneliness.

In addition, there is often a societal lack of understanding of the complexity and severity of the condition, which can lead to stigmatisation and a lack of empathy. This can further impede the social support that is so important for recovery. Economically, Long COVID leads to significant labour absenteeism as many sufferers are unable to start work. This reduces productivity and leads to increased absenteeism, which not only affects individual careers and incomes, but also puts a strain on companies and the economy as a whole. The long-term treatment of Long COVID also significantly increases healthcare costs, which places a financial burden on public and private healthcare systems.

Countermeasures could include flexible working arrangements and adjustments in the workplace that could help to reintegrate those affected into the labour market. Such adaptations could include adjustments to working hours, the option to work from home and special support in the workplace. Furthermore, strengthening social safety nets through special programmes and services tailored to the needs of people with long-term illnesses could provide financial and psychological support for those affected. Public health campaigns and educational programmes are also needed to raise awareness and understanding of Long COVID, helping to reduce stigma and improve social support.

These comprehensive strategies are important to mitigate the impact of Long COVID and manage the burden the disease places on individuals and society. Only through a comprehensive and well-coordinated approach can we hope to effectively address the multiple challenges posed by Long COVID.

Prevention of Long COVID is another key concern in public health policy, especially given the potential long-term impact on individuals and societies worldwide. Future perspectives in prevention include a combination of vaccination strategies, early medical intervention and broader public health measures that are currently being discussed and developed.

Vaccination strategies have proven effective, with studies showing that COVID-19 vaccination can reduce the risk of developing Long COVID, even in those who experience a breakthrough infection after vaccination. Maintaining high vaccination rates, including booster vaccinations, is critical to boosting immunity in the population and limiting the spread of the virus.

Early identification and treatment of COVID-19, especially in at-risk patients, can help prevent progression to severe disease and potentially the development of Long COVID. Implementing systems for early detection of COVID-19 and Long COVID symptoms can help to respond quickly to outbreaks and initiate appropriate treatments.

Promoting healthy lifestyles, including diet, regular physical activity and avoiding risky behaviours, can strengthen the immune system and improve overall health, which in turn can increase resilience to COVID-19 and other viral diseases. In addition, providing resources to manage stress and promote mental health can help improve well-being and mitigate the impact of long-term health problems.

Information campaigns are important to educate the public about the risks and prevention strategies of COVID-19 and Long COVID. Strengthening the health infrastructure to cope with pandemics, including the availability of testing centres, treatment facilities and specialised Long COVID clinics, is also of great importance.

Continuous scientific research and international cooperation are essential to investigate the causes, mechanisms and effective prevention strategies of Long COVID. The development of new technologies to improve the early detection and management of COVID-19, such as innovative testing methods or digital health technologies, play an important role in this.

By implementing this integrated strategy, which includes both individual and public health initiatives, healthcare systems can be strengthened and the social and economic burden of Long COVID can be reduced.

So despite the challenges that Long COVID presents, there has been encouraging progress in the treatment and understanding of these conditions. Researchers around the world have gained significant insights into the mechanisms and effects of the disease, leading to more targeted and effective treatments. The availability and efficacy of COVID-19 vaccines have proven critical in reducing the severity of symptoms and reducing the risk of developing Long COVID. In addition, interdisciplinary approaches that integrate medical, physiotherapeutic and psychological support are continuously improving the quality of life of those affected. This integrative approach reflects a growing recognition and adaptation of healthcare systems to the needs of Long COVID patients and offers hope for those living with the long-term consequences of this global health crisis.